S0-DXN-639

THE STANDARD DATA ENCRYPTION ALGORITHM

THE STANDARD DATA ENCRYPTION ALGORITHM

Harry Katzan, Jr.
CHAIRMAN, COMPUTER SCIENCE DEPARTMENT
PRATT INSTITUTE

a petrocelli book
new york / princeton

Printed in the United States of America

1 2 3 4 5 6 7 8 9 10

Library of Congress Cataloging in Publication Data

Katzan, Harry.
The standard data encryption algorithm.

"A Petrocelli book."
Includes index.
1. Computers–Access control–Passwords. I. Title.
QA76.9.A25K37 651.8 77-13582
ISBN 0-89433-016-0

CONTENTS

PREFACE

Considerations of privacy and confidentiality in a computer environment have given recognition to the need for protecting certain communications and stored data from theft and misuse. The need to protect sensitive data is not unique to the computer field; however, the speed and generality of modern computer systems have compounded the problem to the extent that the traditional means of insuring data security are ineffective. In a computer environment, deliberate or accidental disclosure of classified data can take place even though commonly accepted data security countermeasures, such as access control or the use of hardware/software protection facilities, have been employed. This is precisely the case because stolen data does not leave a theft record and because data protected under hardware or software control is vulnerable without that control.

A suitable methodology for protecting communicated or stored data involves the use of cryptographic techniques. The basic idea is that prior to data transmission, using telecommunications facilities, or data storage on an external medium, such as tape or disk, the data is enciphered—or coded, as it is commonly known. After being enciphered, the data is unintelligible if intercepted or stolen. After transmission or retrieval, the data is subsequently deciphered—or decoded, as it is commonly known—prior to processing.

The national concern for data protection in a modern computer environment resulted in the selection of a standard encryption algorithm by the National Bureau of Standards. The standard data encryption algorithm is the subject of this book, and the presentation includes the following general topics:

1. Introduction to data security and a survey of cryptographic techniques.
2. Description of the standard data encryption algorithm.
3. Detailed implementation of the standard data encryption algorithm in a modern programming language, including computer listings for encipherment and decipherment.

This book is organized so that the reader may select topics of particular interest without necessarily wading through a dearth of detailed material. A reader interested only in the standard encryption algorithm may go directly to the algorithm and then to its implementation. A reader interested additionally in background and supplementary information may select from the other topics, such as data security, survey of cryptographic techniques, detailed walkthrough of the standard data encryption algorithm, and its implementation in a modern programming language.

Two other comments are in order. First, it is not necessary that a given computer installation use the standard data encryption algorithm except as governed by prevailing local regulations. An installation may develop its own algorithm for enciphering/deciphering, and the book contains material that will aid in this respect. Second, the standard data encryption algorithm may be implemented through programming or through special-purpose hardware. In the latter case, the mathematical steps of the algorithm may comprise one or more Large Scale Integration (LSI) chips or a Medium Scale Integration (MSI) electronic package. When used by U.S. federal agencies, the standard data encryption algorithm must be implemented in special-purpose hardware.

It is a pleasure to acknowledge the cooperation of Mr. Orlando Petrocelli, who had the foresight to publish the book, and my wife Margaret, who assisted in preparing the manuscript.

THE STANDARD DATA ENCRYPTION ALGORITHM

INTRODUCTION TO DATA SECURITY

1.1 INTRODUCTION

We live in a computer-based society with a predilection for storing information in centralized files. To some extent, however, this was also the case before the computer era. For example, information on births and deaths has always been stored in a centralized location, such as the county courthouse. Similarly, FBI, Social Security, and Veterans information have likewise been stored in centralized files in the nation's capitol. In older times, physical barriers, such as walls, file cabinets, and ominous well-constructed buildings provided adequate security, since the time, cost, and risk of obtaining information illegally was not commensurate with the value of the information. Computers have not created the data security problem, but the widespread use of computers together with increased demands for information-gathering activities have compounded the problem. With computer-based information systems, it is simply more convenient and more efficient to obtain information—regardless if the access is legal or illegal and whether it is executed on a deliberate or accidental basis. This chapter discusses the data security problem and some of the methods that have been identified for eliminating or minimizing it.

1.2 BASIS OF THE DATA SECURITY PROBLEM

The root of the data security problem lies in an individual's or an organization's need for privacy. The subject is given the most attention when the security of data is compromised, but exists as an integral part of society and influences the behavior of its members. Alan F. Westin defines privacy as:[1] "Privacy is the claim of individuals, groups, or institutions to determine for themselves when, how, and to what extent information about them is communicated to others." The role of privacy in determining an individual's psychological state is well known, but an organization has similar requirements in order to achieve its basic objectives. Much of the information on decision making, competitive products, operational procedures, and on private communications among members must necessarily be kept confidential to insure organizational success.

1.3 DEFINITION OF DATA SECURITY

Communicated or stored information becomes a data security problem when the receiving party does not have the authority to receive or collect it. In a related vein, use of restricted information by an organizational employee for personal gain or for transfer to another person is also a data security problem—even though that individual may have been granted access to the same or similar information for public or organizational purposes.

The subject of data security has been given considerable attention by governmental agencies, computer manufacturers, software companies, and users. Threats to data security have been identified, and effective countermeasures have been established. The IBM Corporation has defined data security as follows:[2] "Data security can be defined as the protection of data from accidental or intentional disclosure to unauthorized persons and from unauthorized modifications." Countermeasures range from a lock on the door to the computer room to the use of cryptographic techniques, and

[1] Alan F. Westin, *Privacy and Freedom* (New York, Atheneum, 1967), p. 7.
[2] "The Considerations of Data Security in a Computer Environment," White Plains, N.Y., IBM Corporation. Form G520-2169, p. 1.

involve removable storage media, such as magnetic disk packs, tape reels, input data, and printed output, as well as the computer facilities. Data security may also involve operational procedures, as summarized in a continuation to the above quotation:[3] "Techniques for security include computer hardware features, programmed routines and manual procedures, as well as the usual physical means of safeguarding the environment with security personnel, locks, keys, and badges."

1.4 REASONS FOR DATA SECURITY

The major reasons that data security is a potential problem in a computer environment stem from the computer itself and the manner in which it is customarily used. First, and possibly most importantly, the illegal access to information in a computer system does not leave a theft record, except when removable storage media is physically stolen, so that it is not always known whether or not the security of a system has been compromised. Second, the resources of a computer system are normally shared among users to make effective use of the equipment. As a result, the special deployment of hardware and software facilities is necessary to provide the required level of data security. Lastly, many modern systems include data communications facilities for transferring information between locations, and a certain amount of exposure to accidental or deliberate access to confidential information exists during the transmission process.

The increased use of telecommunications facilities as a mode of operation in modern computer applications is a secondary reason for the increased concern over data security, and many of the countermeasures pertain to cases in which physical security measures are not sufficient. The problem lies in the fact that in a local operating environment, the computer operator has control over the processing to a greater or lesser extent, whereas in an on-line operating environment, visual means of security are not possible. This fact coupled with the need for varying levels of access to information stored in a centralized data base have markedly increased the scope of data security.

[3] "The Considerations of Data Security in a Computer Environment," *op. cit.*, p. 7.

1.5 THREATS TO DATA SECURITY

Data security countermeasures are varied and complex because the threats to data security may occur in a variety of forms and in many contexts. Overall, threats to data security have been classified as accidental or deliberate. While much of the concern over data security relates to deliberate infiltration, the accidental disclosure of sensitive information can be equally serious.

Accidental

An accidental compromise of data security can result from a hardware failure, a software error, a faulty systems design, or an operational mistake such as mounting the wrong storage medium or entering the wrong information. Regardless of the reason, an accidental penetration of a data management system may make confidential information available to unauthorized persons and provide the opportunity of altering or destroying a file or determining a person's information interests.

Deliberate

The deliberate penetration of a data management system may take place passively or actively. Passive infiltration is similar to wiretapping and involves observing the informational traffic of a system at some point. Passive techniques apply primarily to the use of data communications facilities, but may also involve mundane activity such as inspecting waste containers for computer printouts that were generated accidentally or inappropriately. Active penetration of a data management system involves one of the following overt acts:

1. The use of legitimate access to a system to obtain unauthorized information by browsing through facilities assigned to other persons.
2. Obtaining information illegally by masquerading as another person, through the use of identification acquired by improper means.
3. The use of hardware features, software limitations, or spe-

cially planted entry points* to access restricted information.

4. The use of an open communications channel to obtain information belonging to a user by intercepting messages between the system and the user and by substituting queries pertinent to the infiltrator's needs. (For example, the infiltrator interrupts a message from the user to the system and substitutes his own query. The reply is received by the infiltrator, who returns an error message to the legitimate user.)
5. Physical penetration of the system by theft of removable storage media, by taking over the operation of the system, or through a position associated with the computer center that permits access to the system.

The objective of deliberate penetration of a data management system can also be to obtain confidential information, a person's informational interests, or the opportunity of altering or destroying files.

1.6 DATA SECURITY COUNTERMEASURES

Data security countermeasures include an organized set of procedural, hardware, and software facilities that collectively prevent an unauthorized person from obtaining information from a data management system. Many of the techniques also prevent an unauthorized person from obtaining free use of system resources. Although each data security countermeasure is more effective against some threats than others, there is no single countermeasure that completely eliminates the data security problem. The use of privacy transformations, also popularly known as cryptographic techniques, is most appropriate for safeguarding data from being intelligible by unauthorized parties and is the subject of this book. Collectively, data security countermeasures are conveniently grouped into six broad classes, listed as follows:

1. Access management
2. Processing limitations

*Specially planted entry points, also known as "trap doors," are software features that permit the security system to be bypassed.

3. Auditing and threat monitoring
4. Privacy transformations
5. Integrity management
6. Level of authorization and data file protection

Optimally, a comprehensive data security system would necessarily include techniques from each of the six classes.

Access Management

Access management, also known as "access control" or "user identification," is a set of techniques designed to prevent unauthorized persons from using computer services or accessing data files. Access management techniques are needed when physical identification of the set of users is not feasible or even possible and a relatively large number of operational functions are available to the user when access to the system is obtained. In general, these techniques involve the following concepts:

1. Terminal protection
2. Terminal identification
3. User identification
4. Providing varying levels of service

A terminal can be protected by placing it in a secure location and by employing a "hardwire" connection to the computer. The advantages of this method are obvious within the scope of physical security countermeasures. However, once that physical security is compromised, the computer system offers no security in itself unless other countermeasures are employed. When data communications lines are used instead of a hardwire connection, the computer can be programmed to respond to terminals with a given address code. The computer contains a list of valid address codes and can respond at a given security level to an incoming message that has been prefixed with an address code from the transmitting terminal. An alternate technique is to provide the terminal with the capability of responding to a computer query with a unique identification code. In either case, a terminal hardware failure may render the service unavailable at a given location.

User identification is achieved through something a user knows, by a physical artifact, or by a personal characteristic. In the first category, identification codes and passwords are most frequently used. A prospective user enters an identification code and a password into the computer and is granted or denied access by a validation system implemented in software. One of the disadvantages of identification codes and passwords is that the act of obtaining an unauthorized identification code or password does not leave a theft record so that security can be compromised without anyone explicitly knowing about it. Users also forget or misplace codes and passwords, which leads to the second category. A physical artifact used for computer security is usually taken to be a key or a badge, which can be inserted into a reader to provide identification and to activate the terminal device. The use of physical artifacts is "quick," in the sense that identification is not time consuming, and also offers a theft record so that a lost or stolen key or badge is noticed immediately. Physical characteristics used for computer security include finger, hand, and voice prints. Devices that read physical characteristics are not in widespread use for technical and economic reasons, but represent an area in which future growth is expected.

Access management countermeasures are commonly used in combination to provide varying levels of service. For example, a user of an on-line reservation system may use a physical artifact to gain access to the system and an access code or a password to obtain a given level of service. "Level of service" normally refers to the files that can be accessed and the degree to which records can be added, modified, or deleted. Also, the execution of critical or sensitive programs are commonly restricted to certain classes of personnel. For example, system-related functions may only be available to a user with a classification of systems programmer, while payroll-related functions may only be available to a user with a classification of payroll programmer/analyst.

In general, access management techniques are reasonably effective, but tend to be time consuming from the user's and the computer's point of view. The user must enter a prescribed sequence of codes, passwords, and other identifying data and must respond to queries from the computer. On the computer end, tables of codes, passwords, authorization levels, and other descriptive data must be maintained and various checks and queries must be made at appropriate points in the execution cycle of a computer application.

Processing Limitations

Processing limitations generally refer to the set of hardware/software facilities that control the manner in which work is processed by a computer system and collectively restrict the domain of a user to assigned or authorized areas. Typical examples are the use of memory protection features and virtual memory methods to limit the address space of a program to an assigned area. Other important considerations in this category are volume identification and verification when removable storage media are employed and "erasing" the contents of main storage, direct-access storage, or magnetic tape after the processing of a sensitive job. Procedural restrictions also apply here in the following circumstances:

1. The use of utility programs for copying or modifying data files.
2. The use of dummy test data.
3. Program change and update procedures.
4. Control of manual files.
5. Development and testing of data security routines and the maintenance of associated tables and lists.

The subject of processing limitations is important because the concepts apply to security threats that occur during "non-normal" operations, for which protection is not supplied by the other countermeasures.

Auditing and Threat Monitoring

Auditing and threat monitoring refers to the practice of recording the attempts to violate the security of a computer system or a data file. When a user attempts to access a system resource to which permission has not been granted, one of several actions can be taken:

1. The terminal device can be disconnected or the job can be aborted after a fixed number of unsuccessful attempts.
2. The system can record the violation attempt (i.e., monitor-

ing) but take no explicit action. After an established number of unsuccessful attempts, appropriate personnel can be notified via the operator's console or through a terminal placed in a security officer's office.

The possibility of being monitored also serves as a deterrent, and the technique of auditing and threat monitoring can range from recording attempts to access certain sensitive files to recording all transactions for a given set of users. An average number of violation attempts can be normally expected—either by accident or as a result of demonstration or training procedures. If the number of attempts increases or decreases markedly, then either a concerted effort is being made to penetrate the system or a means of illegally accessing the system has been discovered.

The process of auditing and threat monitoring has an important side benefit. The security log generated through the recording activity within the system can serve as a means of determining how efficiently data files are structured and, as a result, may reveal that a restructuring of a sensitive file into sensitive and nonsensitive components may be desirable.

Privacy Transformations

The term *privacy transformations* refers to the use of cryptographic techniques to conceal the contents of a message or of a data record. In the former case, a message is "encoded" prior to transmission so that information that is intercepted during the transmission process is not immediately intelligible. On the receiving end, the message is "decoded" so that it exists in its original form. The use of techniques and procedures of this type are common in military and governmental affairs and are known as cryptography. In the latter case, data to be stored on an external storage medium is "encoded" prior to being placed on that storage medium so that if an unauthorized person were able to gain access to a data file accidentally or deliberately, the information content would not be readily discernible. When encoded data is read from the external storage medium back into the computer for processing, it is "decoded" into its original form.

The use of privacy transformations provides a relatively high level of security against hardware errors, software errors, wiretapping and the physical theft of storage media. An ultrasophisticated cryptographic technique need not be used. The objective is simply to make the cost of deciphering coded information greater than the value of the information would be to the unauthorized person.

A privacy transformation is a reversible set of logical and arithmetic operations that are performed on the characters of a message or a data record to make the information unintelligible for computer processing or for human recognition. An introduction to cryptographic techniques is contained in chapter 2.

Integrity Management

Integrity management relates to the integrity of hardware, software, people, and operating procedures and to the use of effective physical security measures. In general, this category refers to the following:

1. Controls over hardware and software modifications
2. Secure storage (i.e., safes, vaults, etc.) for programs and data
3. Physical and electrical protection against wiretapping, electromagnetic pickup, and microwave interference
4. Personnel loyalty and integrity
5. Physical controls over admission to the computer area
6. Security procedures for logs and audit facilities
7. Development of secure standard operating procedures for computer operations, security controls, physical transportation of sensitive data, and the restart and recovery of jobs that have failed for hardware or software reasons

Integrity management may also involve the "need to know" with regard to sensitive data and the establishing of organizational policy over the type and amount of information necessary to do a particular job.

Level of Authorization and Data File Protection

The question of *level of authorization* comes up after a user has gained access to a computer system. If the user executes his own programs and accesses his own data files, then the data security problem concerns only access management (covered earlier). However, if programs and data are shared on a systemwide basis, then different levels of authorization are needed to govern the functions that a given user is allowed to perform.

A user's authorization level is established by access management routines during the sign-on (or log-in) procedure. The user's identification code or password serves as an index to a security table that contains an authorized level for various system resources. Then, when a user attempts to utilize a particular resource, a resource manager program is called to verify that the user possesses the correct level of authorization. The following levels give a general idea of sample functions that could be performed at hypothetical levels of authorization:

Level	*Function*
1	Execute a program or read a data file.
2	Modify a program or add a record to a data file, but not delete a statement of the program or a record from the data file.
3	Change a program or data file with additions or deletions.
4	Make a copy of the program or file.
5	Erase the complete program or file.

Authorization may be established for a particular resource on an individual basis, such as "only J. Smith may use this resource," or on a category of users, such as "only systems programmers may use this resource."

Additional protection for data files can be obtained by augmenting the access management and level of authorization techniques through the use of a data file lockword that must be supplied by the program or console operator when a data file is opened for input or output processing. A copy of the lockword is stored with the data file. When the file is opened by a data management routine, the stored lockword is compared with the supplied lockword. Access is permitted only if both entries agree. The technique of using a lockword for data file protection is particularly significant when access management facilities are not otherwise available.

An extension to data file protection allows protection below the file level to individual records or fields. Protection of this type normally requires a data definition facility comprised of a data directory and a decision procedure that is normally stored with the file, so that the owner of the data may specify the criteria needed to access a given element of data.

1.7 IMPLEMENTATION OF DATA SECURITY COUNTERMEASURES

The individual user of a computer system normally has little control over the data security countermeasures that are available, except to the extent that a given set of needs can serve as input in determining an overall data security plan. Data security countermeasures are available primarily through hardware and software features that are usually provided by the computer vendor or developed on an installationwide basis. Privacy transformations are an exception to the rule, since enciphering and deciphering routines can be developed by an individual programmer and can be easily applied to the transmission or storage of data. In fact, many of the less sophisticated methods can be programmed in a relatively short period of time and provide a reasonably high level of data security, depending upon the needs of a particular application.

SELECTED READINGS

Bates, W. S. "Security of Computer-Based Information Systems," *Datamation,* volume 16, number 5 (May 1970), pp. 60-65.

Brown, W. F. *Computer and Software Security.* New York: AMR International, Inc., 1971.

"Considerations of Data Security in a Computer Environment, The" White Plains, N.Y.: IBM Corporation, Form G520-2169.

Hoffman, L. J. "Computers and Privacy: A Survey," *Computing Surveys,* volume 1, number 2 (June 1969), pp. 85-103.

Katzan, H. *Computer Data Security.* New York: Van Nostrand Reinhold Company, 1973.

Miller, Arthur R. *The Assault on Privacy.* Ann Arbor, Michigan: The University of Michigan Press, 1971.

Peterson, H. E. and Turn, R. "System Implications of Information Privacy," *Proceedings of the 1967 Spring Joint Computer Conference,* AFIPS, volume 30, pp. 291–300.
Van Tassel, D. *Computer Security Management.* Englewood Cliffs, N.J.: Prentice-Hall Inc., 1972.
Ware, W. H. "Security and Privacy in Computer Systems," *Proceedings of the 1967 Spring Joint Computer Conference,* AFIPS, volume 30, pp. 279-282.
——"Security and Privacy: Similarities and Differences," *Proceedings of the 1967 Spring Joint Computer Conference,* AFIPS, volume 30, pp. 287–290.
Westin, Alan F. *Privacy and Freedom.* New York: Atheneum, 1967.

2 SURVEY OF CRYPTOGRAPHIC TECHNIQUES FOR DATA PROTECTION

2.1 INTRODUCTION

The terminology privacy "transformation" denotes the use of cryptographic methods to protect certain communicated and stored data against theft or misuse–regardless if the disclosure is legal or illegal and whether it is executed on a deliberate or an accidental basis. Data encryption can be used to supplement other data security countermeasures in order to provide the maximum protection for sensitive data against penetration of the data security system, either actively or passively, by unauthorized persons. The cryptographic technique that should be used in a given case is necessarily dependent upon the nature of the threat to data security. The simplest method of encipherment can prevent accidental disclosure, while the most sophisticated cryptographic technique would provide very little protection against a professional cryptoanalyst. The time and cost required for the utilization of a given method is also of concern, since the process of enciphering and deciphering may or may not be cost effective for the owner and also the infiltrator. There is no *a priori* restriction on how a cryptographic technique should be implemented and both hardware devices and programmed routines have been used effectively.

2.2 BASIC CONCEPTS

Cryptography is defined as the method and process of transforming intelligible test into an unintelligible form and reconverting the unintelligible form into the original text through a reversal of the process of transformation. The transformation process from intelligible text to unintelligible text is known as *enciphering* or *encryption.* The reversal of the process from unintelligible form to original text is known as *deciphering* or *decryption.* The original intelligible text is known as *plain text* or *clear text.* The transformed unintelligible form is known as *cipher text.* The algorithm for performing encryption and decryption is referred to as a *cipher system* that is comprised of one or more of the following methods:

1. Transposition methods
2. Substitution methods
3. Algebraic methods

A *transposition cipher* consists of the rearrangement of the characters in the plain text; the characters retain their identity but lose their position. An elementary example of a transposition cipher is one in which the characters of the plain text are written in reverse order and are transmitted or recorded in fixed-sized groups, such as in the following example:

Plain text: PRIME SUSPECT IS H JONES
Cipher text: SENOJ HSITC EPSUS EMIRP

An alternate transposition cipher involves writing *each* word of the plain text in reverse order and then transmitting or recording the cipher text in fixed-sized groups, as in:

Plain text: IBM WILL SPLIT ON FRIDAY
Cipher text: MBILL IWTIL PSNOY ADIRF

Transposition ciphers are also known as "permutations"; other methods are covered later in this chapter.

A substitution cipher consists of the replacement of characters of the plain text with characters from another alphabet; the characters retain their position but lose their identity. An elementary example of a substitution cipher uses a single cipher alphabet that consists of a plain text component and a cipher text component, as in the following:

Plain text alphabet: A B C D E F G H I J K L M N O P Q R S T U V W X Y Z
Cipher text alphabet: G H I J K L M N O P Q R S T U V W X Y Z A B C D E F

During a transformation operation from plain text to cipher text, the characters comprising the plain text are replaced by the corresponding characters from the cipher text alphabet. The plain text PRIME MINISTER TO ARRIVE TODAY would be enciphered as VXOSK SOTOY ZKXZU GXXOB KZUJG E, where, as before, the cipher text is recorded in fixed-sized groups of characters. Deciphering is essentially the reversal of the enciphering process.

An *algebraic cipher* may involve either of two methods of representation of the plain text and associated transformations as follows:

1. Replacing the plain text characters with numbers using a prearranged conversion and then using the numeric values as input to a reversible series of mathematical operations that produce a numeric result, which is transformed to cipher text through a reversal of the original conversion.
2. Using the bitwise equivalents of the plain text characters according to a suitable binary-coded-decimal coding structure as input to a set of logical and arithmetic operations that produce a binary result, which is transformed back into binary-coded decimal as the cipher text.

Algebraic ciphers lend themselves to automatic computation and are frequently used as one of the methods in a complex cipher system.

2.3 KEYS, MIXED ALPHABETS, AND MIXED NUMBERS

In a substitution cipher, the cipher text alphabet is referred to as a *substitution key,* or simply a *key.* A substitution key can be a standard alphabet, such as the one given above that is arbitrarily established, or a mixed alphabet generated through the use of a key word or a key phrase.

Two methods are commonly employed to mix an alphabet. Both methods use a key word or phrase that can be stored in a restricted table of identification codes, passwords, and keys. The first method is summarized as follows:

1. A key word or phrase, such as UNIVERSITY, is selected.
2. Repeated letters are eliminated, giving UNIVERSTY.
3. The modified key is followed (i.e., suffixed) with the letters of the standard alphabet from which the letters of the key word have been eliminated, as in:

   UNIVERSTYABCDFGHJKLMOPQWXZ

Similarly, the key word METAPHYSICS would be used to generate the mixed alphabet METAPHYSICBDFGJKLNQRUVWXZ and the key word HIPPOCRATES would be used to generate the mixed alphabet HIPOCRATESBDFGJKLMNQUVWXYZ.

The second method uses a matrix of characters to mix an alphabet, as follows:

1. A key word or phrase, such as CLEOPATRA, is selected.
2. Repeated letters are eliminated, giving CLEOPATR.
3. The letters form the first row of a matrix, and successive rows of the matrix are formed from the characters of the standard alphabet from which the letters of the key word have been eliminated, as in:

```
C L E O P A T R
B D F G H I J K
M N Q S U V W X
Y Z
```

4. The characters in the matrix are selected in column order producing the mixed alphabet, i.e.,

   CBMYLDNZEFQOGSPHUAIVTJWRKX

Similarly, the key word PUNISHMENT, together with the resulting matrix

```
P U N I S H M E T
A B C D F G J K L
O Q R V W X Y Z
```

would yield the mixed alphabet PAOUBQNCRIDVSFWHGXMJYEK ZTL.

A *mixed number,* or more accurately, a *set of mixed numbers,* is used to select columns of a matrix of characters according to an established convention. A set of mixed numbers is generated as follows:

1. A key word or phrase, such as ARISTOTLE, is selected.
2. The characters are numbered in accordance with their relative order of appearance in the standard alphabet, numbering repeated letters in sequence from left to right, as in:

   ```
   A R I S T O T L E
   1 6 3 7 8 5 9 4 2
   ```

3. The set of mixed numbers is obtained by selecting the numbers from left to right, i.e., 1 6 3 7 8 5 9 4 2.

Similarly, the key word SHAKESPEARE would yield the following set of mixed numbers: 10 6 1 7 3 11 8 4 2 9 5. Mixed numbers are used with certain transposition and substitution ciphers.

2.4 TRANSPOSITION CIPHERS

Transposition ciphers are rarely used by themselves, since they are relatively easy to decipher because letter frequencies in the plain text are invariant in the cipher form. However, they serve as an important part of an effective cipher system, in combination with other methods. This section covers three transposition methods that are easily implemented on a digital computer. Information on transposition methods that utilize special figures, dials, grilles, or wheels can be found in the literature on cryptography.

Rail-Fence Cipher System

A simple transposition method that dates back to the Civil War is known as the rail-fence cipher system. In one version of this method, the first half of the plain text is written on one line and the second half is written directly below it, as in the following enciphering of the plain text message DIRECTORS CHOOSE PLAN A:

```
D I R E C T O R S C
H O O S E P L A N A
```

The cipher text is generated by selecting the columns from left to right and recording the text in fixed-sized groups as follows:

DHIOR OESCE TPOLR ASNCA

In another version of the rail-fence cipher system, the plain text is placed in columns from left to right, as in:

```
D R C O S H O E L N
I E T R C O S P A A
```

The cipher text is then generated by writing the characters in the two rows in fixed-sized groups, as follows:

DRCOS HOELN IETRC OSPAA

A key is not required with the rail-fence cipher system, which is a minor advantage. However, letter frequencies are invariant in the cipher text, such that the system provides minimal protection when used by itself.

Route Cipher System

A route cipher system involves the placement of the characters of the plain text in a matrix in a prescribed sequence and then the selection of the characters in the matrix using another sequence to generate the cipher text. For example, the message IBM TO PRESS FOR OUT OF COURT SETTLEMENT might be placed in a 6 $\times$ 6 matrix by rows, as follows:

```
I B M T O P
R E S S F O
R O U T O F
C O U R T S
E T T L E M
E N T
```

The cipher text is generated by selecting the characters by column and recording the message in fixed-sized groups of characters as follows:

IRRCE EBEOO TNMSU UTTTS TRLOF OTEPO FSM

Variations to a route cipher system are easily obtained by using other sequences for inscription and selection. Using the same message, variations to the method of inscription might be:

Counterclockwise

```
I B M T O P
O U R T S R
C N T   E E
F E     T S
O M E L T S
T U O R O F
```

Alternating Direction by Row

```
I B M T O P
O F S S E R
R O U T O F
S T R U O C
E T T L E M
E N T
```

Using the following plain text message in matrix form

```
I B M T O P
R E S S F O
R O U T O F
C O U R T S
E T T L E M
E N T
```

variations to the method of transcription (i.e., selection) giving cipher text might be:

Straight-Diagonals Starting at Upper Left Corner: IRBRE MCOST EOUSO ETUTF PNTRO OTLTF ESM

By Columns in Reverse Order: POFSM OFOTE TSTRL MSUUT TBEOO TNIRR CEE

As with the rail-fence system, no key is required with a route cipher system; however, the size of the matrix must be known. This feature adds a degree of protection. When the plain text exceeds the size of the matrix, additional matrices may be utilized. When the characters to be enciphered are less than the size of the matrix, the positions in the matrix can be left blank or null characters, such as the letter X, can be inserted to fill the matrix.

Keyed-Columnar Ciphers

A keyed-columnar transposition cipher uses a key word and a corresponding mixed number. The plain text message is inscribed by rows in a matrix for which each column is numbered by a mixed number, as demonstrated in the encipherment of the following plain text message MAJOR PRODUCT ANNOUNCEMENT FRIDAY AM using the key word ARISTOTLE:

A	*R*	*I*	*S*	*T*	*O*	*T*	*L*	*E*
1	*6*	*3*	*7*	*8*	*5*	*9*	*4*	*2*
M	A	J	O	R	P	R	O	D
U	C	T	A	N	N	O	U	N
C	E	M	E	N	T	F	R	I
D	A	Y	A	M				

The cipher text is transcribed by selecting the columns in numerical order using the mixed number, as follows:

MUCDD NIJTM YOURP NTACE AOAEA RNNMR OF

The last row of the matrix can be filled with null characters, but the security level of the cipher is increased if null characters are not added.

When a transposition cipher is performed once, it is called a *monophase transposition.* Multiple transpositions are referred to as *polyphase transpositions.* The following example enciphers the plain text message BOMBER CONTRACT CANCELLED STOP BID using two keyed-columnar transpositions using different key words:

C O M P U T E R	*S Y S T E M*
1 4 3 5 8 7 2 6	*3 6 4 5 1 2*
B O M B E R C O	B N N T C C
N T R A C T C A	D M R E P O
N C E L L E D S	T C O B A L
T O P B I D	B O A S R T
	E D E C L I

The cipher text from the leftmost keyed-columnar transposition is selected in numerical order by column from the first set of mixed numbers and is inscribed in the rightmost matrix by row. The final cipher text is transcribed in numerical order by column from the second set of mixed numbers, generating the following cipher text:

CPARL COLTI BDTBE NROAE TEBSC NMCOD

When deciphering a keyed-columnar transposition, the length of the key word is used to determine the width of the matrix. The length of the cipher text is divided by the length of the key word; the quotient gives the number of full rows in the matrix and the remainder gives the number of characters in the last and possibly incomplete row. Once the dimensions of the matrix are established, the columns can be filled vertically using the set of mixed numbers determined by the key word.

2.5 SUBSTITUTION CIPHERS

As mentioned above, a substitution cipher uses two alphabets: a plain text alphabet and a cipher text alphabet. The plain text alphabet corresponds to the characters of the plain text message and the cipher text alphabet contains the respective cipher text equivalents. Substitution ciphers are classified according to the following attributes:

Attribute	*Discussion*
Monoalphabetic vs. Polyalphabetic	Refers to the number of alphabets that are employed during the substitution
Uniliteral vs. Multiliteral	Refers to the number of cipher characters that replace a single character from the plain text
Monographic vs. Digraphic	Refers to the number of characters in the plain text that are processed at one time

All variations of the above attributes are not covered here; however, the number of distinct options indicate the power inherent in substitution ciphers.

Monoalphabetic/Uniliteral/Monographic Substitution

In this type of substitution cipher, one alphabet is used to replace one character of the plain text with a single character from the cipher alphabet. The single substitution cipher given in section 2.2, Basic Concepts, is an example of a cipher of this type. A mixed alphabet is frequently used with monoalphabetic/uniliteral/monographic, as demonstrated in the following example that enciphers the plain text PRIME MINISTER TO ARRIVE SUNDAY with the key word PUNISHMENT using a columnar transposition to mix the alphabet (as covered in section 2.3):

Plain text alphabet:
A B C D E F G H I J K L M N O P Q R S T U V W X Y Z
Cipher text alphabet:
P A O U B Q N C R I D V S F W H G X M J Y E K Z T L

The resultant cipher text message is:

HXRSB SRFRM JBXJW PXXRE BMYFU PT

Another option is to use a reciprocal alphabet that facilitates the enciphering/deciphering process. In a reciprocal alphabet, the same algorithm and alphabets are used for both operations, since, for example, A in plain text enciphers into an L in cipher text and L in plain text enciphers into an A in cipher text. The following is a reciprocal alphabet

Plain text alphabet:
A B C D E F G H I J K L M N O P Q R S T U V W X Y Z
Cipher text alphabet:
L K J I H G F F D C B A Z Y X W V U T S R Q P O N M

where INSTITUTE in plain text enciphers into DYTSDSRSH in cipher text, and DYTSDSRSH in plain text enciphers into INSTITUTE in cipher text.

A third variation is to mix both alphabets, as follows:

Plain text alphabet:
A R I S T O L E B C D F G H J K M N P Q U V W X Y Z
Cipher text alphabet:
V W X Z M E T A P H Y S I C B D F G J K L N O Q R U

The two alphabets are viewed as two character sequences being slid against each other, such that A(plain) = V(cipher). In this case, 26 slides are possible and the first letter of the cipher text is used as an indicator of the particular slide being used. Thus, for example, the plain text message CONCESSION EXPECTED TODAY would be enciphered as

VHEGH AZZXE GAQJA HMAYM EYVR

where the initial V denotes the particular slide of the cipher text alphabet that is used.

Monoalphabetic/Multiliteral/Monographic Substitution

In this type of substitution cipher, one alphabet is used and two or more characters of the cipher text alphabet are substituted for each character in the plain text message. A straightforward version of this method uses a key word inscribed in a substitution matrix, and row and column indices are labeled by letters as follows:

		A	B	C	D	E
	A	U	N	IJ	V	E
First Letter	B	R	S	T	Y	A
	C	B	C	D	F	G
	D	H	K	L	M	O
	E	P	Q	W	X	Z

The row/column indices, taken as pairs, are used to represent the characters in the substitution matrix. Thus, for example, the plain text character Y is represented as BD; plain text character I as AC; plain text character P as EA, and so forth. (The plain text characters I and J are used interchangeably to obtain a square matrix.) Using a monoalphabetic/multiliteral/monographic substitution and the above substitution matrix, the plain text message REVERT TO PLAN XMAS would be enciphered as:

BAAEA DAEBA BCBCD EEADC BEABE DDDBE BB

The substitution matrix for monoalphabetic/multiliteral/monographic substitution can be formed in other ways. A common variation is to generate a mixed alphabet, such as the following,

C L E O P A T R
B D F G H IJ K M
N Q S U V W X Y
Z

and inscribe it in the substitution matrix by rows as follows:

	A	E	I	O	U
A	C	B	N	Z	L
E	D	Q	E	F	S
I	O	G	U	P	H
O	V	A	IJ	W	T
U	K	X	R	M	Y

and, as shown, some variation is also possible in choosing the row and column indices.

Polyalphabetic/Uniliteral/Monographic Substitution

A well-known polyalphabetic/uniliteral/monographic substitution cipher is known as the Vigenère Cipher that uses the Vigenère Table, given in Table 2.1. The Vigenère Cipher operates as follows:

1. A selected key word or key phrase is written above the plain text message.
2. Each plain text character is enciphered through the use of

the Vigenère Table. The cipher text equivalent of a plain text character is the character located at the intersection of the column headed by the plain text character and the row identified by the corresponding character from the key. For example, the plain text character M enciphered with key character P yields the cipher text character B through the use of the Vigenère Table. In other words, $M_{plain} (P_{key}) \rightarrow B_{cipher}$ and $G_{plain} (T_{key}) \rightarrow Z_{cipher}$.

Table 2.1 The Vigenère table used with the Vigenère cipher

Plain Text Character

Key Character	A	B	C	D	E	F	G	H	I	J	K	L	M	N	O	P	Q	R	S	T	U	V	W	X	Y	Z
A	A	B	C	D	E	F	G	H	I	J	K	L	M	N	O	P	Q	R	S	T	U	V	W	X	Y	Z
B	B	C	D	E	F	G	H	I	J	K	L	M	N	O	P	Q	R	S	T	U	V	W	X	Y	Z	A
C	C	D	E	F	G	H	I	J	K	L	M	N	O	P	Q	R	S	T	U	V	W	X	Y	Z	A	B
D	D	E	F	G	H	I	J	K	L	M	N	O	P	Q	R	S	T	U	V	W	X	Y	Z	A	B	C
E	E	F	G	H	I	J	K	L	M	N	O	P	Q	R	S	T	U	V	W	X	Y	Z	A	B	C	D
F	F	G	H	I	J	K	L	M	N	O	P	Q	R	S	T	U	V	W	X	Y	Z	A	B	C	D	E
G	G	H	I	J	K	L	M	N	O	P	Q	R	S	T	U	V	W	X	Y	Z	A	B	C	D	E	F
H	H	I	J	K	L	M	N	O	P	Q	R	S	T	U	V	W	X	Y	Z	A	B	C	D	E	F	G
I	I	J	K	L	M	N	O	P	Q	R	S	T	U	V	W	X	Y	Z	A	B	C	D	E	F	G	H
J	J	K	L	M	N	O	P	Q	R	S	T	U	V	W	X	Y	Z	A	B	C	D	E	F	G	H	I
K	K	L	M	N	O	P	Q	R	S	T	U	V	W	X	Y	Z	A	B	C	D	E	F	G	H	I	J
L	L	M	N	O	P	Q	R	S	T	U	V	W	X	Y	Z	A	B	C	D	E	F	G	H	I	J	K
M	M	N	O	P	Q	R	S	T	U	V	W	X	Y	Z	A	B	C	D	E	F	G	H	I	J	K	L
N	N	O	P	Q	R	S	T	U	V	W	X	Y	Z	A	B	C	D	E	F	G	H	I	J	K	L	M
O	O	P	Q	R	S	T	U	V	W	X	Y	Z	A	B	C	D	E	F	G	H	I	J	K	L	M	N
P	P	Q	R	S	T	U	V	W	X	Y	Z	A	B	C	D	E	F	G	H	I	J	K	L	M	N	O
Q	Q	R	S	T	U	V	W	X	Y	Z	A	B	C	D	E	F	G	H	I	J	K	L	M	N	O	P
R	R	S	T	U	V	W	X	Y	Z	A	B	C	D	E	F	G	H	I	J	K	L	M	N	O	P	Q
S	S	T	U	V	W	X	Y	Z	A	B	C	D	E	F	G	H	I	J	K	L	M	N	O	P	Q	R
T	T	U	V	W	X	Y	Z	A	B	C	D	E	F	G	H	I	J	K	L	M	N	O	P	Q	R	S
U	U	V	W	X	Y	Z	A	B	C	D	E	F	G	H	I	J	K	L	M	N	O	P	Q	R	S	T
V	V	W	X	Y	Z	A	B	C	D	E	F	G	H	I	J	K	L	M	N	O	P	Q	R	S	T	U
W	W	X	Y	Z	A	B	C	D	E	F	G	H	I	J	K	L	M	N	O	P	Q	R	S	T	U	V
X	X	Y	Z	A	B	C	D	E	F	G	H	I	J	K	L	M	N	O	P	Q	R	S	T	U	V	W
Y	Y	Z	A	B	C	D	E	F	G	H	I	J	K	L	M	N	O	P	Q	R	S	T	U	V	W	X
Z	Z	A	B	C	D	E	F	G	H	I	J	K	L	M	N	O	P	Q	R	S	T	U	V	W	X	Y

The Vigenère Cipher is normally used by repeating the key word, as in

Key word:
C A T A S T R O P H E C A T A S T R O P H E C A T A
Plain text:
D I C T A T O R D E A D C O U P I N P R O G R E S S
Cipher text:
F I V T S M F F S L E F C H U K B E D G V K T E L S

which is known as *periodic polyalphabetic substitution,* or by using a nonrepeating key phrase, as in

Key phrase:
N O W I S T H E T I M E F O R A L L G O O D M E N T
Plain text:
D I C T A T O R D E A D C O U P I N P R O G R E S S
Cipher text:
Q W Y B S M V V W M M H H C L P T Y V F C J D I F L

The latter case is known as *aperiodic polyalphabetic substitution,* and a well-known phrase or a line of a book is agreed upon in advance as the key phrase.

Monoalphabetic/Uniliteral/Digraphic Substitution

In a substitution cipher of this type, two characters from the plain text are processed at one time. One of the better-known *digraphic substitution* methods is the Playfair Cipher that uses a substitution matrix, based on the use of a key word as follows:

O U T L A
W B C D E
F G H IJ K
M N P Q R
S V X Y Z

The Playfair Cipher is used as follows:

1. The plain text message is divided into groups of two characters. Pairs of identical characters are separated by an infrequently used character, such as X or Z. The message

THE VIENNA AFFAIR, for example, is divided as:

TH EV IE NZ NA AF FA IR

In the matrix, each pair of characters will be in the same row, same column, or neither.

2. If a pair of plain text characters are in the same row, the corresponding cipher text characters are the characters to the immediate right of the plain text characters; e.g., $(WC)_{plain} \rightarrow (BD)_{cipher}$ and $(NR)_{plain} \rightarrow (PM)_{cipher}$
3. If a pair of plain text characters are in the same column, the corresponding cipher text characters are the characters immediately below the plain text characters; e.g., $(TH)_{plain} \rightarrow (CP)_{cipher}$ and $(BV)_{plain} \rightarrow (GU)_{cipher}$
4. If a pair of plain text characters is neither in the same row nor the same column, the cipher text equivalent of each character is the character found at the intersection of its own row and the column of the other character; e.g., $(EV)_{plain} \rightarrow (BZ)_{cipher}$ and $(IE)_{plain} \rightarrow (KD)_{cipher}$

Through the use of the Playfair Cipher, the message THE VIENNA AFFAIR is enciphered as

CP BZ KD RV RU OK KO KQ

and transcribed as CPBZK DRVRU OKKOK Q

A large number of substitution ciphers have been developed and the interested reader is directed to *The Code-Breakers* by David Kahn[1] for additional information on the subject.

2.6 ALGEBRAIC SYSTEMS

As mentioned previously, algebraic ciphers are usually defined on sequences of binary digits or numeric equivalents of plain text characters. After a table look-up operation to perform the character-to-numeric translation, enciphering and deciphering are performed through the use of conventional mathematical methods.

[1] D. Kahn, *The Code-Breakers* (New York: The Macmillan Company, 1967).

The Vernam Cipher

The Vernam Cipher was invented for use with teletype code and uses the exclusive-OR operation, defined as follows:

	Exclusive OR	Plain Text 1	Plain Text 0
Key	1	0	1
	0	1	0

Thus, for example, if the plain text were 010001 and the key were 110111, then the encipherment is:

Plain text:	0 1 0 0 0 1
Key:	1 1 0 1 1 1
Cipher text:	1 0 0 1 1 0

The Vernam Cipher is convenient since encipherment and decipherment are reciprocal.

Practical use of the Vernam Cipher requires a table of BCD codes, such as the one included as Table 2.2. A simple procedure for using the Vernam Cipher is given as follows:

1. Convert the plain text and the key to binary codes using the conversion table.
2. Perform the Vernam Cipher yielding the cipher text in binary form.
3. Convert the binary cipher text to character form using the conversion tables.

For example, the plain text message ATTACK using the key word HOTDOG would be enciphered as follows using the Vernam Cipher:

Binary plain text:
010001 110011 110011 010001 010011 100010

Binary key:
011000 100110 110011 010100 100110 010111

Binary cipher text:
001001 010101 000000 000101 110101 110101

Table 2.2 Representative conversion table for 6-bit BCD codes

Character	*Octal*	*Binary*
0	00	000000
1	01	000001
2	02	000010
3	03	000011
4	04	000100
5	05	000101
6	06	000110
7	07	000111
8	10	001000
9	11	001001
#	12	001010
@	13	001011
?	14	001100
:	15	001101
>	16	001110
⩾	17	001111
+	20	010000
A	21	010001
B	22	010010
C	23	010011
D	24	010100
E	25	010101
F	26	010110
G	27	010111
H	30	011000
I	31	011001
.	32	011010
[	33	011011
&	34	011100
(	35	011101
<	36	011110
←	37	011111
X	40	100000
J	41	100001
K	42	100010
L	43	100011
M	44	100100
N	45	100101
O	46	100110

Table 2.2 Continued

Character	*Octal*	*Binary*
P	47	100111
Q	50	101000
R	51	101001
$	52	101010
*	53	101011
-	54	101100
)	55	101101
;	56	101110
⩽	57	101111
Blank	60	110000
/	61	110001
S	62	110010
T	63	110011
U	64	110100
V	65	110101
W	66	110110
X	67	110111
Y	70	111000
Z	71	111001
'	72	111010
%	73	111011
≠	74	111100
=	75	111101
]	76	111110
"	77	111111

and the resulting cipher text would be 9E05VV after it was converted from binary to character form.

The intermediate binary form of the cipher text lends itself to permutation operations prior to being converted to character form.

Cipher System Based on Simultaneous Equations

A cipher system based on the use of simultaneous equations was developed by Lester S. Hill.[2] In the following equations, the x's represent plain text characters and the y's represent cipher text

[2] Kahn, *op. cit.*, p. 405.

characters, according to the following arbitrarily established alphabet:

A	B	C	D	E	F	G	H	I	J	K	L	M	N	O	P	Q	R	S	T	U	V	W
4	8	25	2	9	20	16	5	17	3	0	22	13	24	6	21	15	23	19	12	7	11	18

X	Y	Z
1	14	10

The cipher, known as a *tetragraphic substitution* because the cipher sequence is based on groups of four plain text characters, involves the use of the following enciphering equations:

$$y_1 = 8x_1 + 6x_2 + 9x_3 + 5x_4$$
$$y_2 = 6x_1 + 9x_2 + 5x_3 + 10x_4$$
$$y_3 = 5x_1 + 8x_2 + 4x_3 + 9x_4$$
$$y_4 = 10x_1 + 6x_2 + 11x_3 + 4x_4$$

The deciphering equations, which must be based on the ciphering equations, are given as follows:

$$x_1 = 23y_1 + 20y_2 + 5y_3 + 1y_4$$
$$x_2 = 2y_1 + 11y_2 + 18y_3 + 1y_4$$
$$x_3 = 2y_1 + 20y_2 + 6y_3 + 25y_4$$
$$x_4 = 25y_1 + 2y_2 + 22y_3 + 25y_4$$

The arithmetic is performed modulo 26. An implementation of the cipher system based on simultaneous equations in the APL language is given in Figure 2.1. The enciphering function is named ENCIPHER and the deciphering function is named DECIPHER. In the demonstrated example, the plain text message HELP is first translated into the following set of numbers:

$$x_1 = 5$$
$$x_2 = 9$$
$$x_3 = 22$$
$$x_4 = 21$$

After encipherment through the use of enciphering equations, the

```
      ∇ENCIPHER[⎕]∇
    ∇ R←ENCIPHER T;⎕IO;I;A;B;X
[1]   ⎕IO←0
[2]   A←'KXDJAHOUBEZVTMYQGIWSFPLRNC'
[3]   B← 4 4 ρ 8 6 9 5 6 9 5 10 5 8 4 9 10 6 11 4
[4]   X←A⍳T
[5]   R←,A[26|B+.×X]
    ∇

      ∇DECIPHER[⎕]∇
    ∇ R←DECIPHER T;⎕IO;I;A;B;Y
[1]   ⎕IO←0
[2]   A←'KXDJAHOUBEZVTMYQGIWSFPLRNC'
[3]   B← 4 4 ρ 23 20 5 1 2 11 18 1 2 20 6 25 25 2 22 25
[4]   Y←A⍳T
[5]   R←,A[26|B+.×Y]
    ∇

      ENCIPHER 'HELP'
UQZY
      DECIPHER 'UQZY'
HELP
```

Figure 2.1 APL implementation of a cipher system based on simultaneous equations

following values for the y's are derived:

$y_1 = 7$

$y_2 = 15$

$y_3 = 10$

$y_4 = 14$

which are translated into the cipher text message UQZY. The calculations are essentially the same in the deciphering process except that the simultaneous equations are replaced to perform decipherment, as covered above.

Cipher System Based on Matrix Methods

The use of simultaneous equations was simplified by Hill[3] to include matrices and matrix algebra in which the concept of an "involutory transformation" was introduced. With an *involutory*

[3] Kahn, *op. cit.*, p. 406.

transformation, the same equations are used for encipherment and decipherment. The matrix equations developed by Hill are given as follows, with a constant matrix added to increase security:

$$Y_1 = \begin{pmatrix} 3 & 6 & 2 \\ 16 & 23 & 8 \\ 2 & 16 & 13 \end{pmatrix} X_1 + \begin{pmatrix} 2 & 6 & 14 \\ 8 & 24 & 4 \\ 14 & 16 & 20 \end{pmatrix} X_2 + \begin{pmatrix} 18 & 6 & 6 \\ 24 & 20 & 22 \\ 2 & 2 & 16 \end{pmatrix}$$

$$Y_2 = \begin{pmatrix} 18 & 14 & 22 \\ 20 & 4 & 10 \\ 22 & 20 & 24 \end{pmatrix} X_1 + \begin{pmatrix} 15 & 16 & 20 \\ 4 & 13 & 2 \\ 20 & 8 & 11 \end{pmatrix} X_2 + \begin{pmatrix} 2 & 16 & 14 \\ 8 & 12 & 4 \\ 18 & 8 & 20 \end{pmatrix}$$

Using the following arbitrary alphabet

A	B	C	D	E	F	G	H	I	J	K	L	M	N	O	P	Q	R	S	T	U	V
4	8	25	2	9	20	16	5	17	3	0	22	13	24	6	21	15	23	19	12	7	11

W	X	Y	Z
18	1	14	10

the plain text message AIR SEA ATTACK AT DAWN is inscribed in matrices and converted to numeric values as follows:

$$X_1 = \begin{pmatrix} A & I & R \\ S & E & A \\ A & T & T \end{pmatrix} = \begin{pmatrix} 4 & 17 & 23 \\ 19 & 9 & 4 \\ 4 & 12 & 12 \end{pmatrix}$$

$$X_2 = \begin{pmatrix} A & C & K \\ A & T & D \\ A & W & N \end{pmatrix} = \begin{pmatrix} 4 & 25 & 0 \\ 4 & 12 & 2 \\ 4 & 18 & 24 \end{pmatrix}$$

The matrix arithmetic, performed modulo 26, generate the following values for Y_1 and Y_2:

$$Y_1 = \begin{pmatrix} 6 & 15 & 3 \\ 25 & 11 & 20 \\ 20 & 16 & 14 \end{pmatrix} = \begin{pmatrix} O & Q & J \\ C & V & F \\ F & G & Y \end{pmatrix}$$

$$Y_2 = \begin{pmatrix} 8 & 1 & 12 \\ 20 & 20 & 24 \\ 10 & 6 & 4 \end{pmatrix} = \begin{pmatrix} B & X & T \\ F & F & N \\ Z & O & A \end{pmatrix}$$

which yields the cipher text message:

OQJC VFFG YBXT FFNZ OA

An APL function for the involutory enciphering and deciphering is given in Figure 2.2 and named HILL.

```
        ∇HILL[⎕]∇
      ∇ R←F HILL T;⎕IO;A;I;X1;X2;Y1;Y2;A1;A2;B1;B2;C1;C2
[1]    ⍝ F=1 DENOTES ENCIPHER, F≠1 DENOTES DECIPHER
[2]     ⎕IO←0
[3]     A←'KXDJAHOUBEZVTMYQGIWSFPLRNC'
[4]     A1← 3 3 ρ 3 6 2 16 23 8 2 16 13
[5]     B1← 3 3 ρ 2 6 14 8 24 4 14 16 20
[6]     C1← 3 3 ρ 18 6 6 24 20 22 2 2 16
[7]     A2← 3 3 ρ 18 14 22 20 4 10 22 20 24
[8]     B2← 3 3 ρ 15 16 20 4 13 2 20 8 11
[9]     C2← 3 3 ρ 2 16 14 8 12 4 18 8 20
[10]    I←A⍳T
[11]    X1← 3 3 ρ9↑I
[12]    X2← 3 3 ρ9↓I
[13]    →(F=1)/ENCPHR
[14]    X1←26|X1-C1
[15]    X2←26|X2-C2
[16]    Y1←26|(A1+.×X1)+(B1+.×X2)
[17]    Y2←26|(A2+.×X1)+(B2+.×X2)
[18]    →FINI
[19]  ENCPHR:Y1←26|(A1+.×X1)+(B1+.×X2)+C1
[20]    Y2←26|(A2+.×X1)+(B2+.×X2)+C2
[21]  FINI:R←(,A[Y1]),,A[Y2]
      ∇


        ⎕←TXT←1 HILL 'AIRSEAATTACKATDAWN'
OQJCVFFGYBXTFFNZOA
        2 HILL TXT
AIRSEAATTACKATDAWN
```

Figure 2.2 APL function for involutory enciphering and deciphering

2.7 SYSTEMS CONCEPTS

When cryptographic techniques are used with stored data, data records are simply enciphered before storage and deciphered after retrieval. When cryptographic techniques are used with communicated data, then the task of providing secure communications is a systems problem as well as a data security problem.

Types of Communications Systems

Two types of communications systems can be readily identified: computer-to-computer and user-to-computer. In a computer-to-computer system, a hardware-implemented cryptographic device can exist for each communications line, as suggested in Figure 2.3(a), or a single hardware device or computer program can service each computer system, as suggested in Figure 2.3(b). The sending and receiving units synchronize the key used for enciphering and deciphering by means of information contained in the header message transmitted.

In a user-to-computer system, software facilities for enciphering and deciphering are not normally available to the user and a hardware-implemented cryptographic device is considered to be the only viable alternative for secure communications. At the computer end, a hardware device or a computer program can be used, as with a computer-to-computer system. A user-to-computer system is suggested in Figure 2.4.

Coordination and Integrity

Coordination refers to the process of establishing keys and passwords at each end of the communications line, and it takes place at two different times: (1) when the system is established and (2) when a message is transmitted. *Integrity* refers to the method of verifying passwords and keys.

In a computer-to-computer system, coordination is usually performed prior to transmission through a sign-on process similar to that employed with user-to-computer systems (covered below). Once coordination is achieved, the enciphering and deciphering facilities are primed and messages are automatically enciphered on the sending end and deciphered on the receiving end. A variation to this concept will be covered later under Computer Networks.

In a user-to-computer system, two methods are identified that are independent of the cryptographic process. In the first method, known as *handshaking,* the user first signs on to the system so that the system may determine the user's cryptographic key from a

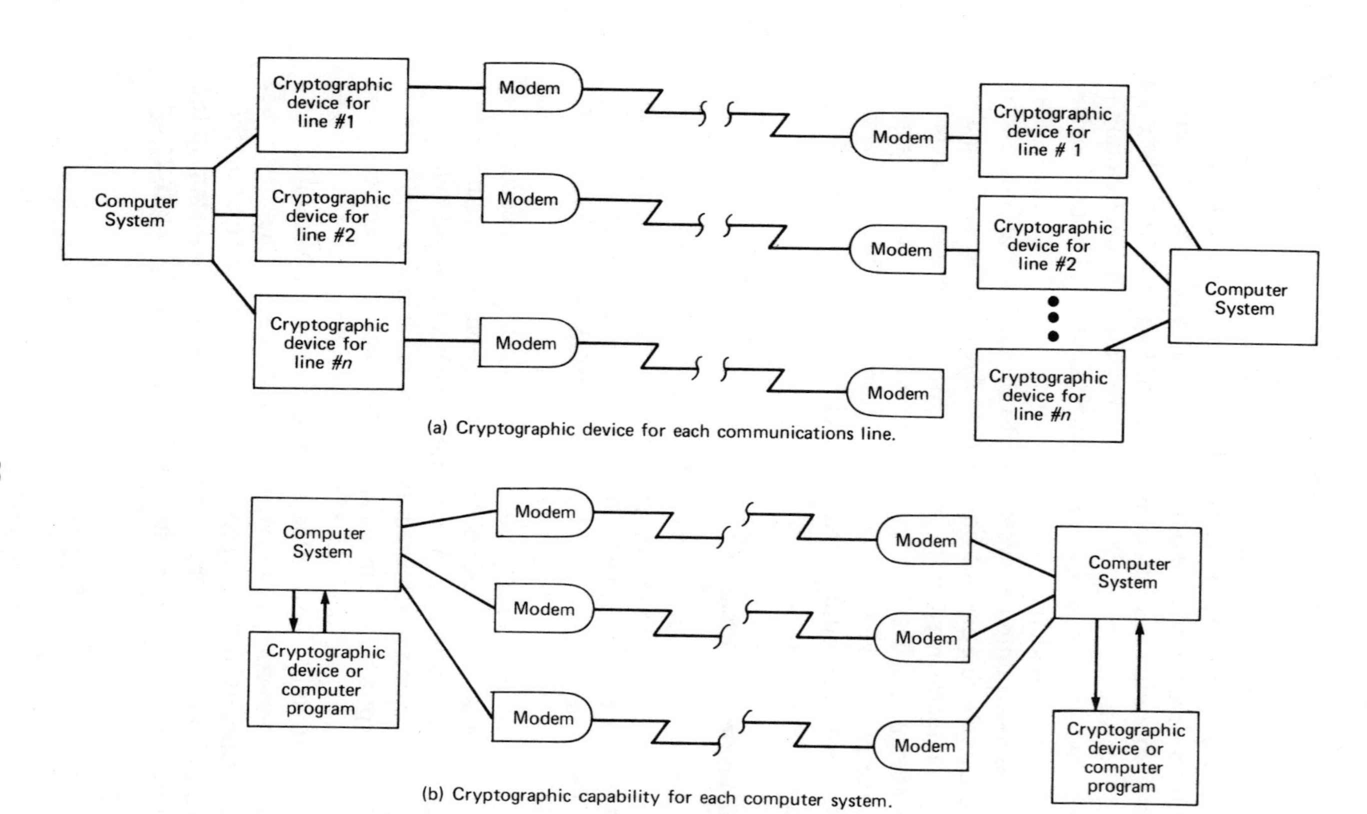

(a) Cryptographic device for each communications line.

(b) Cryptographic capability for each computer system.

Figure 2.3 Types of computer-to-computer communications systems: (a) cryptographic device for each communications line; (b) cryptographic capability for each computer system

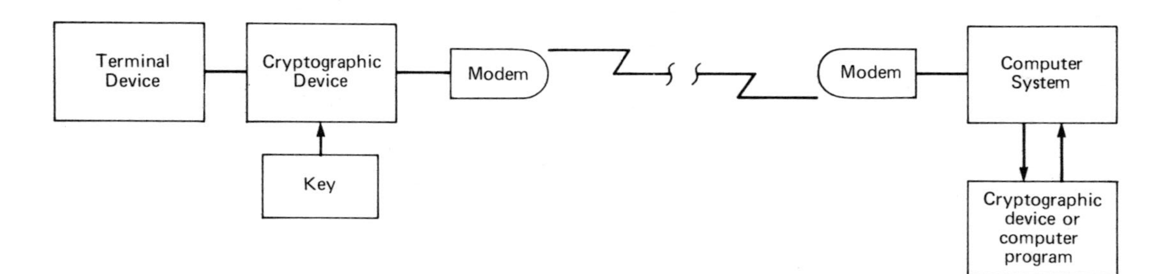

Figure 2.4 A user-to-computer system

secure table at the computer site. A series of transactions are initiated when the user states his identity (U) in plain text and transmits it along with an arbitrary message (M) enciphered with his key. The computer uses U's key to decipher the arbitrary message. The computer then appends its own arbitrary message (N) to M, enciphers it using U's key, and transmits it back to U. When U receives the message, a comparison of message M, as sent and as received, verifies the computer's identity to U. On the next transmission from the user to the computer, the computer's message N is appended to the user's message so that the computer can verify the user's identity. A handshaking procedure of this type protects against a "between the lines" infiltration of the system. The second method uses a predetermined password—for each user—that is included in the data block being transmitted in enciphered form. Again, the user must first sign-on to the system so that the same cipher key can be used. The computer uses the user's cipher key, obtained from a directory as covered above, to decipher the message and compare passwords. Other methods for achieving coordination are covered in the data security literature and involve enciphering messages in steps wherein the total message may involve the transmission of several blocks in enciphered form.

In a secure operating environment, cipher keys are assigned with user identification codes and passwords. Depending upon the required level of security, cipher keys must be changed at regular intervals because large samples of information facilitate the cryptographic process. When a change of cipher keys is made, it should be transported between locations by carrier or by registered mail. In user-to-computer systems, enciphering and deciphering may be performed at the user's end by a hardware device that uses a magnetically encoded card, similar to a credit card, with the user's cipher key. The use of a magnetically encoded card permits the use of a standard enciphering/deciphering algorithm wherein the cipher system is based only on the key and not on the method.

Computer Networks

A computer network is characterized by several computer systems and end users communicating together in a communications mode. In a system of this type, link encipherment or an end-to-end encipherment can be used. With *link encipherment,* enciphering/deciphering devices and/or software facilities are placed functionally at the modem interfaces. All data is transmitted in enciphered form and the enciphering/deciphering process is essentially transparent to the sending and receiving stations. Link encipherment is satisfactory for wiretapping but does not protect against a message being misrouted, either accidentally or intentionally by a penetrator. With *end-to-end encipherment,* information is enciphered at its source and is not deciphered until it reaches its destination. Thus, it is not significant if a message is misrouted because it is of very little value to any station except the intended destination. A typical end-to-end encipherment message format is given in Figure 2.5. The network header and link control information cannot be enciphered, but the information content of the message is enciphered.

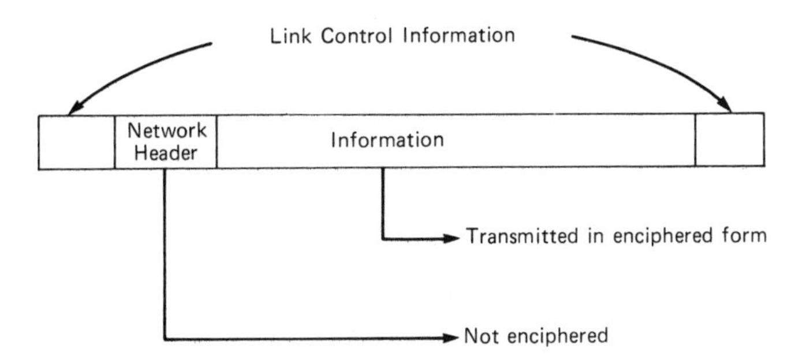

Figure 2.5 Message format for *end-to-end encipherment*

SELECTED READINGS

AFIPS, volume 42, *Proceedings of the 1973 National Computer Conference*: Mellon, G. E. "Cryptotology, Computers, and Common Sense," pp. 569–579; Meyer, C. H. "Design Considerations for Cryptography," pp. 603–606; Reed, I. S. "Information Theory and Privacy in Data Banks," pp. 581-587; Stahl, F. A. "A Homophonic Cipher for Computational Cryptography," pp. 565–568; Turn, R. "Privacy Transformations for Databank Systems," pp. 589-601.

Kahn, D. *The Code Breakers.* New York: The Macmillan Company, 1967.

Katzan, H. *Computer Data Security.* New York: Van Nostrand Reinhold Company, 1973.

Krishnomurthy, E. V. "Computer Cryptographic Techniques for Processing and Storage of Confidential Information," *International Journal of Control,* volume 12, number 5 (1970), pp. 753-761.

Smith, J. L., Notz, W. A., and Ossick, P. R. "An Experimental Application of Cryptography to a Remotely Accessed Data System," *Proceedings of the ACM Annual Conference,* August, 1972, pp. 282-297.

Sykes, David J. "Protecting Data by Encryption," *Datamation,* volume 22, number 8 (August 1976), pp. 81-85.

Van Tassel, D. "Advanced Cryptographic Techniques for Computers," *Communications of the ACM,* volume 12, number 12 (December 1969), pp. 664-665.

3 OVERVIEW OF THE DATA ENCRYPTION STANDARD

3.1 INTRODUCTION

The *Data Encryption Standard* (DES) is a mathematical algorithm to be used for the cryptographic protection of computer data. The algorithm is designed for use with binary-coded data and uses a 64-bit key to encipher (i.e., encript) 64 bits of information. The 64-bit key is of prime importance since a unique key results in the cryptographic generation of a unique set of 64 bits of cipher text from 64 bits of plain text. Since the algorithm is in general known by all persons involved with the data encryption standard, the cryptographic security of the information is dependent upon the security used to protect the key. Encripted information can be transformed into the original plain text through a reversal of the algorithmic process using the same key that was employed for encryption.

The data encryption algorithm was designed so that 56 bits of the 64-bit key are used for the encryption process and the remaining 8 bits are used only as parity error-detecting bits. More specifically, the key is divided into eight 8-bit bytes. In an 8-bit byte, 7 bits are used by the algorithm and the eighth bit can be used to maintain odd parity. From a complete 64-bit block of plain text enciphered with a 56-bit key, there is no known technique, other than trying all possible keys, for determining the cipher key. With

56 bits, there are over 70,000,000,000,000,000 (i.e., seventy quadrillion) possible combinations so that the chances of determining a specific key through an exhaustive search is extremely unlikely. In fact, Dr. Ruth Davis, Director of the National Bureau of Standard's Institute for Computer Science and Technology, in speaking about the standard encryption algorithm has stated, "that while no code is 'theoretically unbreakable,' 2500 years of computer time on a general-purpose computer 'significantly' faster than a CDC 7600 would be required to derive a key." And, Dr. Davis estimated that from matched sets of clear and cipher data, "well over $100 million could be spent five years from now to find a key . . . ."[1]

3.2 RATIONALE FOR THE DATA ENCRYPTION STANDARD

The rationale behind the data encryption standard is that certain communicated and stored data can have significant value or sensitivity and the need for adequate protection has become a national issue.[2] Although data security countermeasures have indeed been developed, it is generally felt that they are more effective against the amateur than the professional. This point of view is summarized well by James R. Kitchen:[3]

> The amateurs or radicals may be deterred by a well-developed bulwark such as solid doors, card locks and security personnel with guard dog. However, professional thieves will consider them routine and proceed with the appropriate counteraction.
>
> These professionals are capable of bypassing alarm systems, finding and tapping the facility's data links, bribing and coercing employees and of writing programs which can bypass software restraints. It would be no problem for a group with this level of expertise to breach a facility and remove, copy or exchange data files.

[1] *Computer,* volume 10, number 2 (February, 1977), p. 6.
[2] *Federal Register,* volume 40, number 149 (Friday, August 1, 1975), p. 32395.
[3] James R. Kitchen, "Cryptography Urged as Practical Means of Security," *Computerworld,* January 17, 1977, p. 18.

In the same article, Kitchen identifies two methods of safeguarding data. The first method is to store the valuable data in a secure location, such as a bank vault; the second method is to employ cryptography.

Since it is generally recognized that cryptography is an effective countermeasure, provided that the encryption algorithm is sufficiently complex,[4] the National Bureau of Standards (NBS) solicited submissions for a standard algorithm to be used by federal agencies for protecting unclassified computer data. The algorithm selected as the standard was submitted by the International Business Machines Corporation, which has granted nonexclusive, royalty-free licenses as governed by patent law to make, use, and sell apparatus that complies with the standard.[5]

3.3 USING THE DATA ENCRYPTION STANDARD

The data encryption standard is intended for use when cryptographic protection is required by federal departments and agencies and by nonfederal government agencies when the needed level of security is provided through the use of the algorithm. More specifically, the standard must be used by federal departments and agencies, subject to the following conditions:

1. The authorized person who is responsible for data security decides that cryptographic protection is required.
2. The data to be enciphered is not classified according to the National Security Act of 1947, as amended, or the Atomic Energy Act of 1954, as amended.

Federal agencies or departments that otherwise use cryptographic techniques may use those techniques in lieu of the standard.

---

[4] *Federal Register, op. cit.,* p. 32395.
[5] *Data Encryption Standard,* FIPS Publication 46, U. S. Department of Commerce/ National Bureau of Standards (January 15, 1977), p. 3.

3.4 IMPLEMENTATION OF THE ALGORITHM

The following quotation outlines the position of the National Bureau of Standards with regard to the implementation of the data encryption standard:[6]

> The algorithm specified in this standard is to be implemented in computer or related data communication devices using hardware (not software) technology. The specific implementation may depend on several factors such as the application, the environment, the technology used, etc. Implementations which comply with this standard include Large Scale Integration (LSI) "chips" in individual electronic packages, devices built from Medium Scale Integration (MSI) electronic components, or other electronic devices dedicated to performing the operations of the algorithm. Micro-processors using Read Only Memory (ROM) or micro-programmed devices using microcode for hardware level control instructions are examples of the latter.

Moreover, the National Bureau of Standards will provide procedures for testing and validating appropriate hardware equipment and will test and validate equipment as complying with the standard. As far as the NBS is concerned, software implementations on general-purpose computers are not in compliance with the standard. However, the standard "applies" to all Federal ADP systems and associated telecommunications networks . . ."[7] but does not necessarily apply to private and nonfederal organizations, except as governed by policy and local regulations. Therefore, implementation of the data encryption standard (DES) algorithm through software on a general-purpose computer is a possible alternative to the acquisition and practical implementation of a hardware device that performs the same function, when an organization's needs do not justify the hardware approach* because the volume of data to be enciphered is small. A major consideration here involves how the encipherment is to be used. If it is to be used for transmitted data,

---

[6] *Data Encryption Standard, op. cit,* p. 2.
[7] *Data Encryption Standard, op. cit,* p. 3.
*It is also possible for an organization to develop its own encryption algorithm through the use of the methods given in chapter 2.

then a hardware device may indeed be the most secure and convenient method of implementation—regardless of cost. If encipherment is to be used only with stored data, then the use of a separate hardware device may be cumbersome and a software encryption scheme—possibly employing methods other than the standard—may be appropriate.

Another consideration is the fact that the instruction set in most modern computers is implemented through the use of microprogramming, and the corresponding microcode is stored in read-only memory (ROM) that is a physical extension to main storage.* Moreover, programmed read-only memory (PROM) is used with most microprocessors and the machine-language instructions stored in and executed out of PROM are the same ones used in conventional assembly-level programming.

Therefore, a reasonable conclusion is that the distinction between a microcoded implementation of the data encryption standard algorithm, which is permitted by the standard,† and a software version is one of degree and that a software version is indeed possible and may have distinct advantages over hardware for some users. Also, a software version may lead to an implementation that complies with the standard once it is written to PROM and validated.

Obviously, there are other reasons behind the hardware level implementation of the algorithm, such as testing, validity, and protection against modification. The point has been made, however; it is possible for nonfederal organizations to effectively utilize a software implementation of the data encryption standard, when particular requirements are such that a hardware implementation is not feasible.

3.5 POSSIBLE MODES OF OPERATION

Two different modes are identified for using the algorithm:

1. *Electronic Code Book Mode.* Data is encrypted using a key in blocks of 64 bits. Each block of plain text and cipher

*Other names for main storage are random-access memory (RAM), in the area of microprocessors, and S-memory, in the area of microprogramming.

†Microcoded implementations, in order to satisfy the requirements of the standard, must of course be validated.

text is independent of preceding and succeeding blocks.

2. *Cipher Feedback Mode.* The algorithm is used as a binary stream generator to produce random bits that are combined with binary plain text using the exclusive or logical operation to form binary cipher text. Input to the algorithm is the previous 64 bits of data that were transmitted or received, and as in the alternate mode, a key must also be used.

In either case, the data encryption algorithm is used in precisely the same manner.

3.6 OVERVIEW OF THE ALGORITHM

The data encryption standard algorithm incorporates the following steps to encipher a 64-bit block of data using a 64-bit key:

1. A transposition operation, referred to as the *initial permutation* (IP). This transposition does not utilize the 64-bit key and operates solely on the 64 data bits.
2. A complex key-dependent *product transformation* that uses block ciphering to increase the number of substitutional and reordering patterns.
3. A final transposition operation, referred to as the *inverse initial permutation* (IP^{-1}), which is an actual reversal of the transformation performed in the first step.

The three major steps in the data encryption standard algorithm are summarized in Figure 3.1.

The initial permutation and the inverse initial permutation are simple bit transpositions; they are covered later. The product transformation requires a more comprehensive introduction.

In cryptography, a *product transformation* is the successive application of substitution and transposition ciphers. When a computer is used, large blocks of data can be transformed as a unit, providing the advantage, mentioned above, of increasing the number of substitutional and reordering patterns. The latter technique is known as *block ciphering.*

In the product transformation step of the data encryption

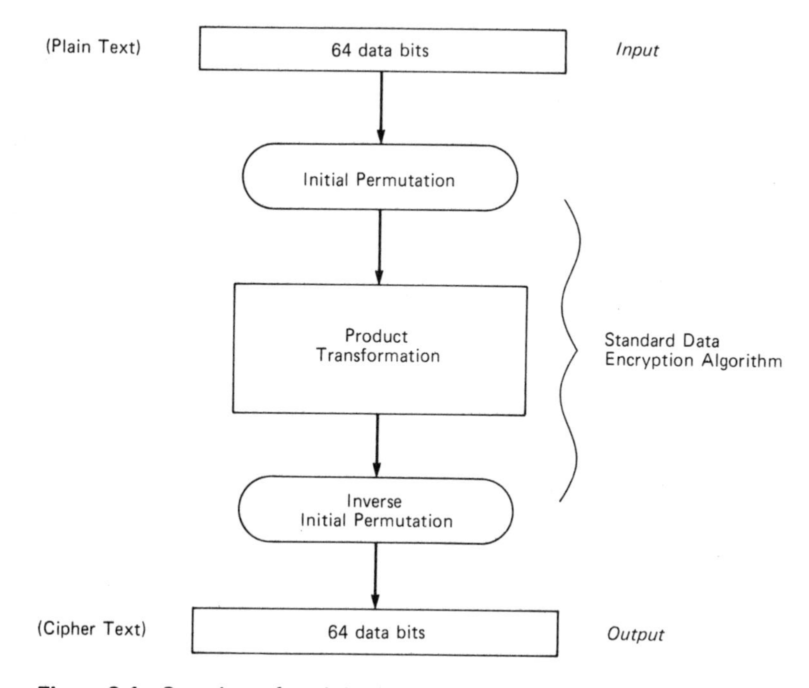

Figure 3.1 Overview of enciphering process for the standard data encryption algorithm

standard algorithm, which is a block ciphering system, substitutions are performed under the control of a cipher key while transpositions are performed according to a fixed sequence. Figure 3.2 depicts one iteration of the product transformation, which includes the following operations:

1. The 64-bit block of plain text is divided into two 32-bit blocks, denoted by L_i and R_i for "left" and "right," respectively.
2. The rightmost 32 bits of the input block become the leftmost 32 bits of the output block. This is denoted in Figure 3.2 by an arrow going from R_{i-1} to L_i.
3. The rightmost 32 bits of the input block, denoted hereafter as R_{i-1}, go through a selection process yielding a 48-bit data block. This is a fixed selection that is not key dependent.

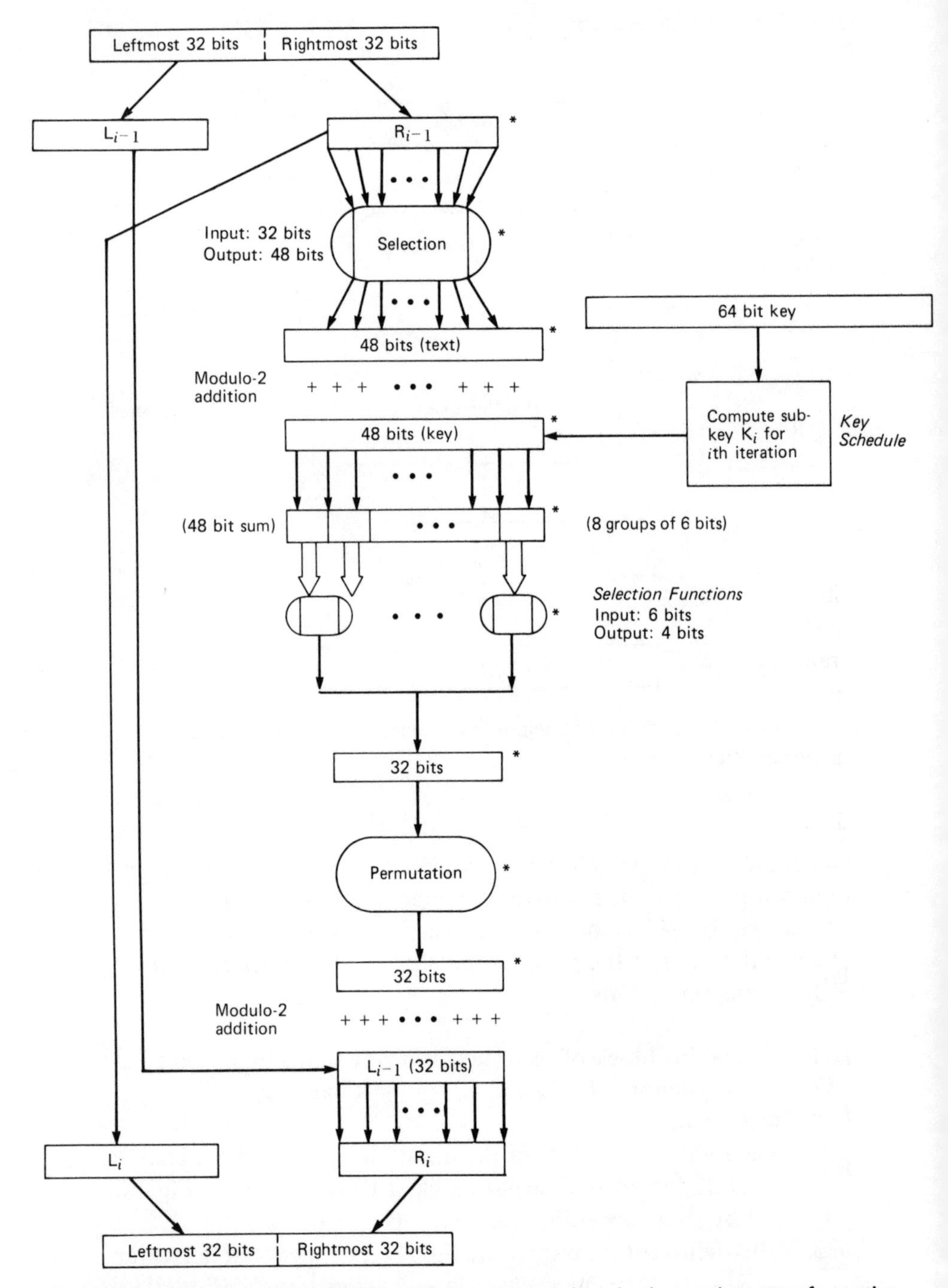

Figure 3.2 Overview of one iteration of the computations in the product transformation of the standard data encryption algorithm (* denotes steps in the cipher function, covered in chapter 4)

4. The 64-bit key is used to generate a 48-bit subkey K_n, where $1 \leqslant n \leqslant 16$. Each K_i is unique and corresponds to the ith iteration of the product transformation.
5. The 48-bit subkey is added (modulo-2) to the output of step 3 yielding a 48-bit result.
6. The 48-bit output of step 5 is divided into eight 6-bit groups, that are each subjected to a unique substitution cipher that yields eight 4-bit groups, that are concatenated to form a 32-bit output.
7. The 32-bit output of step 6 is permuted to produce a 32-bit block. (This is a simple transposition.)
8. The 32-bit output of step 7 is added (modulo-2) to the leftmost 32 bits of the input block (denoted by L_{i-1}) yielding R_i, which is the rightmost 32 bits of the 64-bit output block.

Steps 1 through 8 are repeated 16 times; this constitutes the major part of the product transformation. The last step in the product transformation is the block transformation (i.e., exchange) of the left and right halves of the output of the last iteration.

The deciphering process is the exact reversal of encipherment, using the subkeys of the product transformation in reverse order, i.e., K_{16} to K_1. Chapter 4 gives a detailed analysis of the steps in the data encryption standard algorithm.

SELECTED READINGS

Bright, H. S., and Enison, R. L. "Cryptography Using Modular Software Elements," *Proceedings of the 1976 National Computer Conference,* AFIPS Volume 45, pp. 113–123.

Data Encryption Standard, FIPS Publication 46, U.S. Department of Commerce/National Bureau of Standards, 1977.

Federal Register, Volume 40, Number 149 (Friday, August 1, 1975), p. 32395ff.

Heinrich, F. R., and Kaufman, D. J. "A Centralized Approach to Computer Network Security," *Proceedings of the 1976 National Computer Conference,* AFIPS Volume 45, pp. 85-90.

Ingemarsson, I. "Analysis of Secret Functions with Application to Computer Cryptography," *Proceedings of the 1976 National Computer Conference,* AFIPS Volume 45, pp. 125–127.

Katzan, H. *Computer Data Security.* New York: Van Nostrand Reinhold Company, 1973.

Kitchen, J. R. "Cryptography Urged as Practical Means of Security," *Computerworld,* January 17, 1977, p. 18.

4 DETAILED ANALYSIS OF THE STANDARD DATA ENCRYPTION ALGORITHM

4.1 COMPONENTS OF THE ALGORITHM

One means of describing the complex standard data encryption algorithm is to present its major components in detailed form and then to cover the ciphering and deciphering processes that incorporate the various components. This approach involves the dissection of the product transformation, introduced in chapter 3, and the identification of the major computational operations contained therein.

Collectively, the following components are described:

1. The *key schedule* calculations, which is a procedure that generates the 16 subkeys
2. The *modulo-2 addition* operation
3. The *cipher function,* which comprises the main operations in the product transformation
4. The *block transposition* that yields a "preoutput block," which serves as input to the inverse initial permutation
5. The *initial permutation* described as a selection table
6. The *inverse initial permutation* described as a selection table

The various components are subsequently written in symbolic form in describing the enciphering and deciphering processes, and as easy-

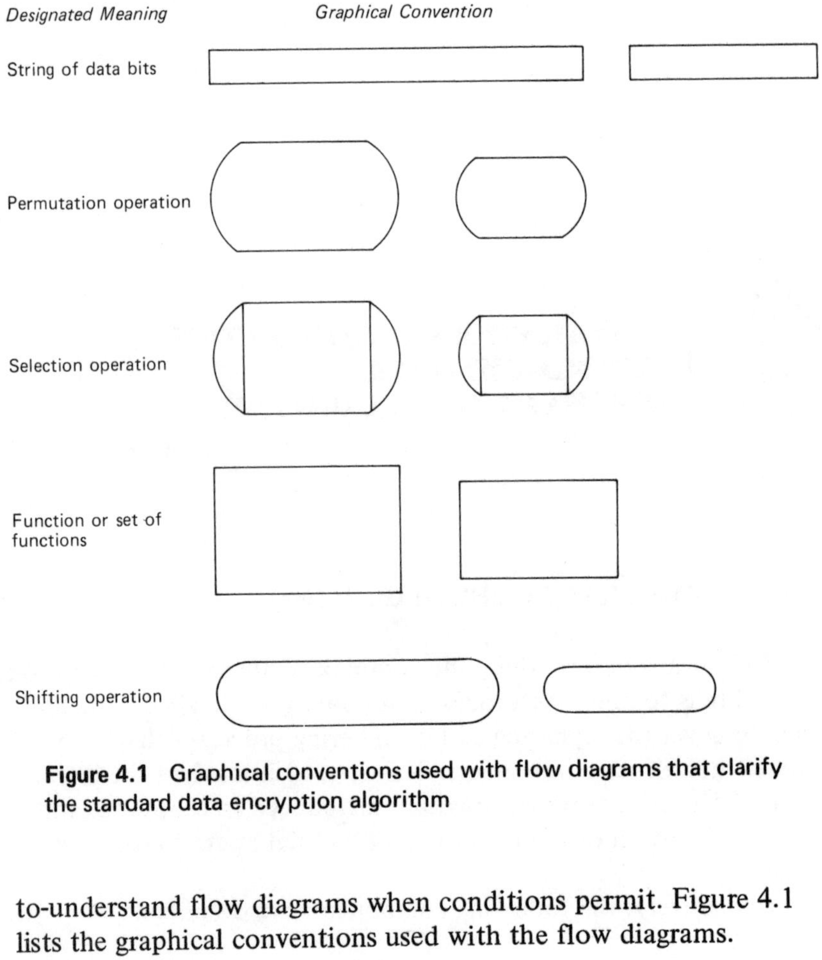

Designated Meaning *Graphical Convention*

String of data bits

Permutation operation

Selection operation

Function or set of functions

Shifting operation

Figure 4.1 Graphical conventions used with flow diagrams that clarify the standard data encryption algorithm

to-understand flow diagrams when conditions permit. Figure 4.1 lists the graphical conventions used with the flow diagrams.

4.2 THE KEY SCHEDULE CALCULATIONS

The objective of the set of key schedule calculations is to generate the 16 subkeys, referred to as K_n, required for the enciphering and deciphering processes. Each K_n is 48 bits long and is derived through the use of permutation, selection, and shifting operations.

The bits in the 64-bit key are numbers from 1 to 64, going from left to right. However, all bits of the key are not used in the key schedule calculations. Recognizing also that the 64-bit key represents eight 8-bit bytes, one bit from each byte is used to maintain

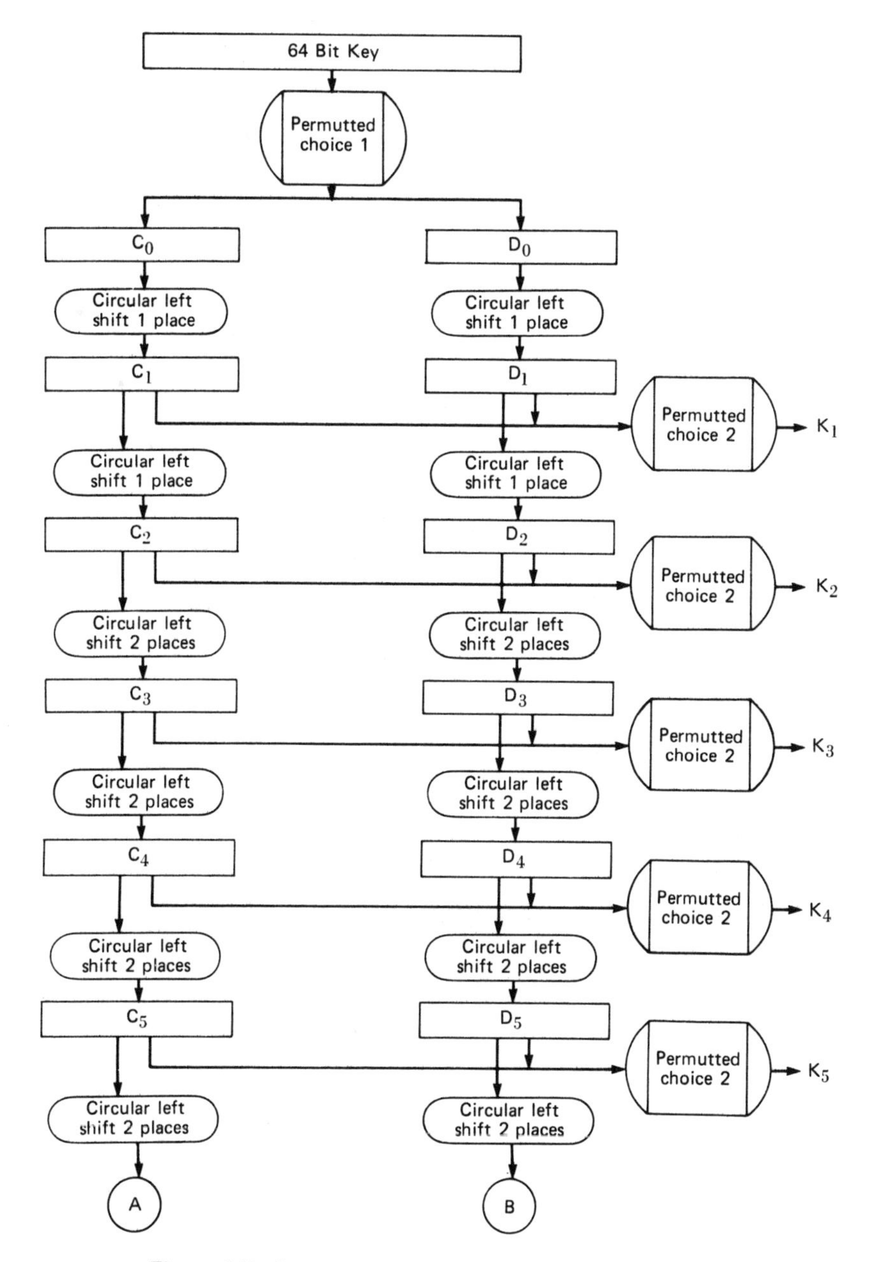

Figure 4.2 Summary of the key schedule calculations

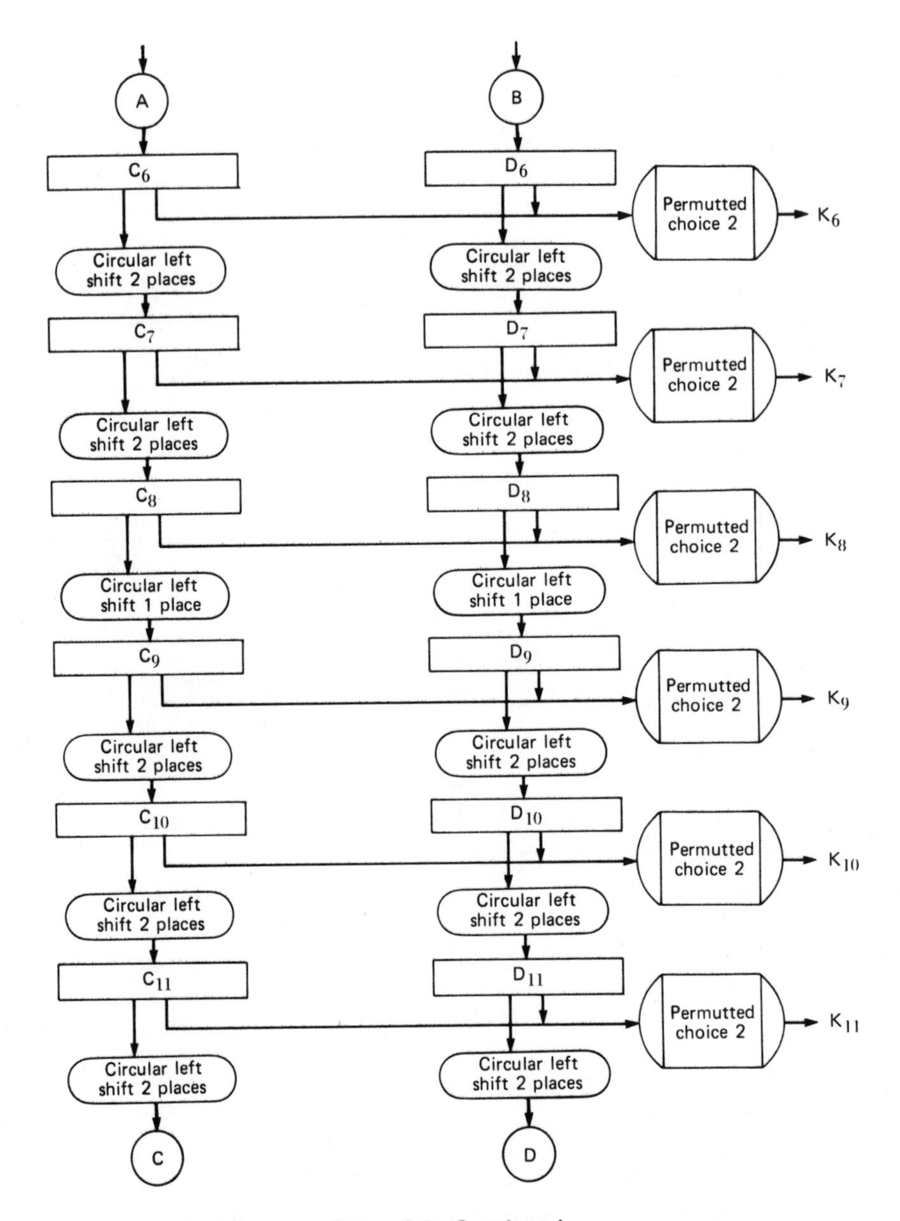

Figure 4.2 Continued

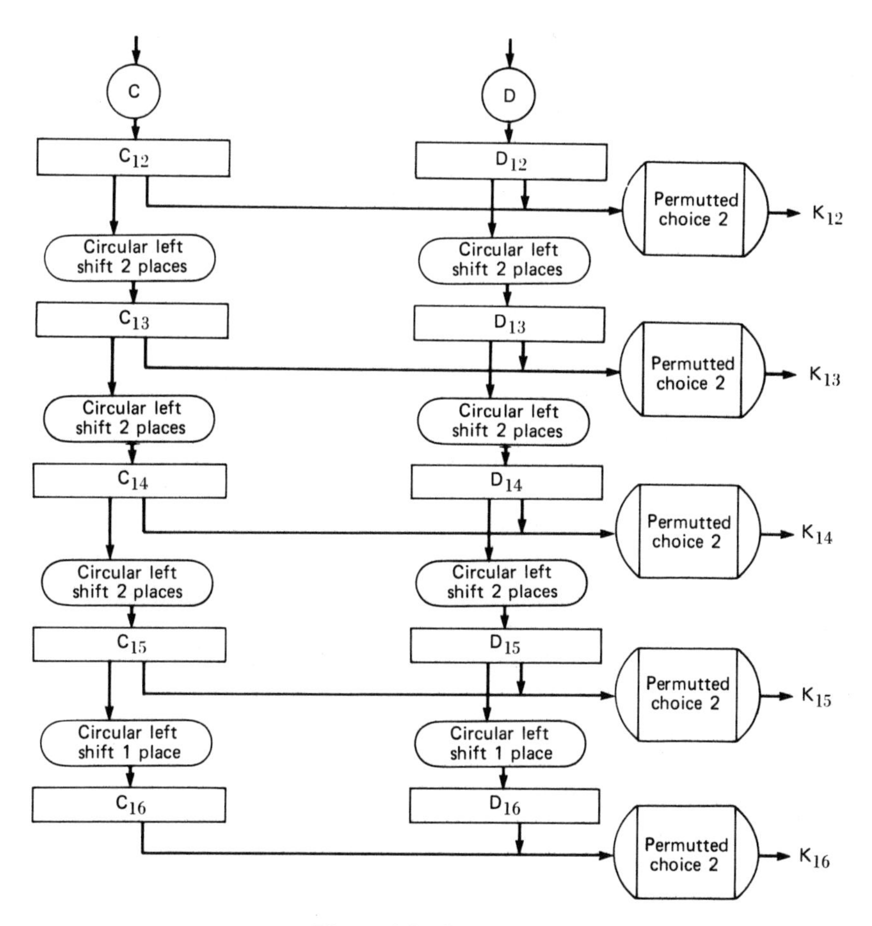

Figure 4.2 Continued

odd parity and is not used in the key schedule calculations.* The parity bits are numbered 8, 16, 24, 32, 40, 48, 56, and 64, leaving the following bits for key schedule computations:

1 through 7
9 through 15
17 through 23
25 through 31
33 through 39
41 through 47
49 through 55
57 through 63

The key schedule computations are executed as follows:

1. The non-parity bits in the key go through a permutation operation yielding two 28-bit blocks denoted by C_0 and D_0. This is the starting point for computing the subkeys.
2. C_0 and D_0 are circularly left shifted one place yielding C_1 and D_1.
3. Selected bits from C_1 and D_1 are tapped off yielding subkey K_1.
4. C_1 and D_1 are circularly left shifted one place yielding C_2 and D_2.
5. Selected bits from C_2 and D_2 are tapped off yielding subkey K_2.
6. The process continues for subkeys K_3 through K_{16}. Each C_i and D_i is obtained from the preceding value after a prescribed number of circular left shifts.

The key schedule calculations are summarized in Figure 4.2. Each subkey, denoted by K_i, is obtained through a selection operation from C_i and D_i. C_i and D_i are obtained from C_{i-1} and D_{i-1}, respectively, through prescribed shift operations.

Initially C_0 and D_0 are obtained from the 64-bit key through the use of *permuted choice 1,* which is summarized by the permutation table in Figure 4.3. Recalling that the bits in the cipher key are numbered from 1 to 64, going from left to right, permuted

*A given implementation of the standard data encryption standard can verify the parity of each byte prior to enciphering or deciphering. (In the sample program, written in APL and given later in the book, parity is not checked.)

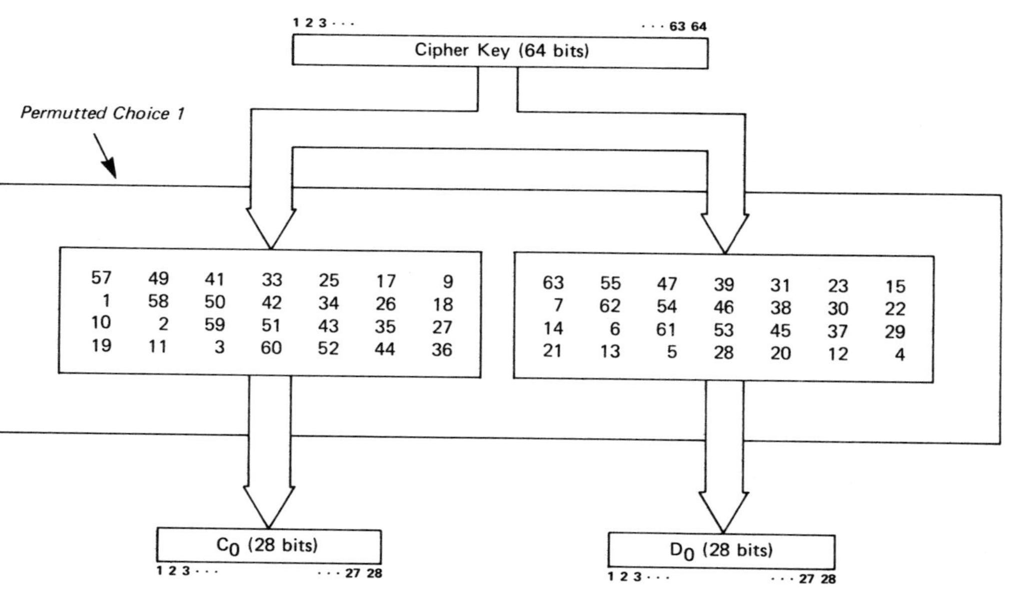

Figure 4.3 Permuted choice 1, used in the calculation of C_0 and D_0

choice 1 specifies that the bits of C_0 are, respectively, bits 57, 49, 41, . . . , 9, 1, 58, 50, . . . , 18, 10, 2, 59, . . . , 27, 19, 11, . . . , 36 of the cipher key. Similarly, the bits of D_0 are respectively bits 63, 55, 47, . . . , 15, 7, 62, 54, . . . , 22, 14, 6, 61, . . . , 29, 21, 13, 5, . . . , 4 of the cipher key.

Permuted choice 2 is used to select a particular key K_n from the concatenation of C_n and D_n. C_n and D_n are each 28 bits long so that C_nD_n combined has bits that run from 1 through 56. Permutated choice 2 is summarized by the selection table in Figure 4.4. The bits in subkey K_n are numbered from 1 to 48; going from left to right, the bits of K_n are, respectively, bits 14, 17, . . . , 5, 3, 28, . . . , 10, 23, 19, . . . , 8, . . . , 46, 42, . . . , 32, of C_nD_n. It should be noted that permutation choice 2 is used in the computation of each of the subkeys K_1 through K_{16}.

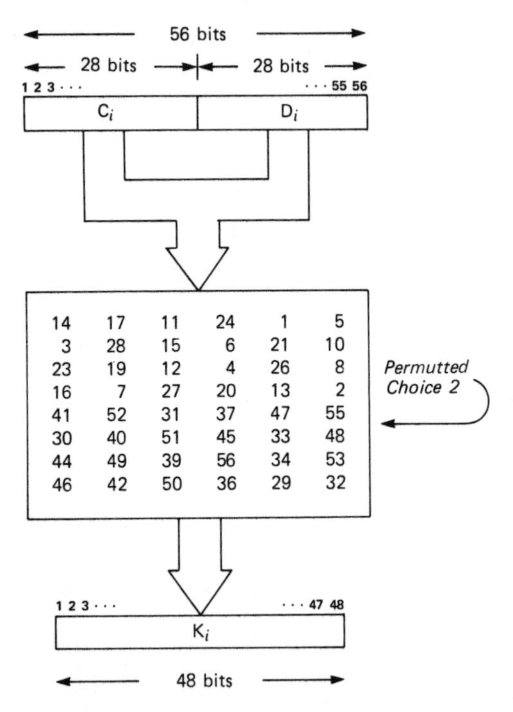

14	17	11	24	1	5
3	28	15	6	21	10
23	19	12	4	26	8
16	7	27	20	13	2
41	52	31	37	47	55
30	40	51	45	33	48
44	49	39	56	34	53
46	42	50	36	29	32

Figure 4.4 Permuted choice 2, used in the calculation of subkey K_i

In the implementation of the standard data encryption algorithm, it is anticipated that the key schedule computations will be computed as an iterative procedure, wherein iteration-1 generates K_1, iteration 2 generates K_2, and so forth. The number of circular left shifts for each iteration of the calculation is summarized in Table 4.1.

Table 4.1 The number of circular left shifts for each iteration in the key schedule calculation

Iteration #	*# of circular left shifts*
1	1
2	1
3	2
4	2
5	2
6	2
7	2
8	2
9	1
10	2
11	2
12	2
13	2
14	2
15	2
16	1

4.3 MODULO-2 ADDITION

A bit-by-bit modulo-2 addition operation is used in several steps of the standard data encryption algorithm. The operation denoted by $\oplus$ is defined as follows:

$\oplus$	0	1
0	0	1
1	1	0

So that the following example would be valid:

$$\begin{array}{r} 1\ 0\ 0\ 1\ 0\ 1\ 1\ 0 \\ \oplus\ 1\ 1\ 0\ 1\ 0\ 0\ 1\ 1 \\ \hline 0\ 1\ 0\ 0\ 0\ 1\ 0\ 1 \end{array}$$

A bit-by-bit addition modulo 2 is the same as the exclusive-OR operation, defined in Boolean algebra as:

$$(X \wedge \overline{Y}) \vee (\overline{X} \wedge Y)$$

where X and Y are the arguments to the exclusive-OR operation.

4.4 THE CIPHER FUNCTION

The *cipher function* comprises the main operations in the product transformation, and the steps in the cipher function are identified with an asterisk in the overview diagram of Figure 3.2. The cipher function is used in each iteration of the product transformation and is specified symbolically as:

$$f(A, K_n)$$

where A is a string of 32 data bits representing R_i for encryption and L_i for decryption, and K_n is a 48-bit subkey determined from the key schedule.

Figure 4.5 gives an overview of the cipher function, which combines the following operations:

1. A selection operation E that operates on the argument A of 32-bits and produces a 48-bit result.
2. A modulo-2 addition which adds the result of the selection operation E and the 48-bit key K_n on a bit-by-bit basis yielding a 48-bit result.
3. A unique set of selection functions S_i that converts the 48-bit result of the modulo-2 addition to a set of 32 bits.
4. A permutation operation P that operates on the 32-bit result of the previous operation (i.e., step 3) and produces a 32-bit result.

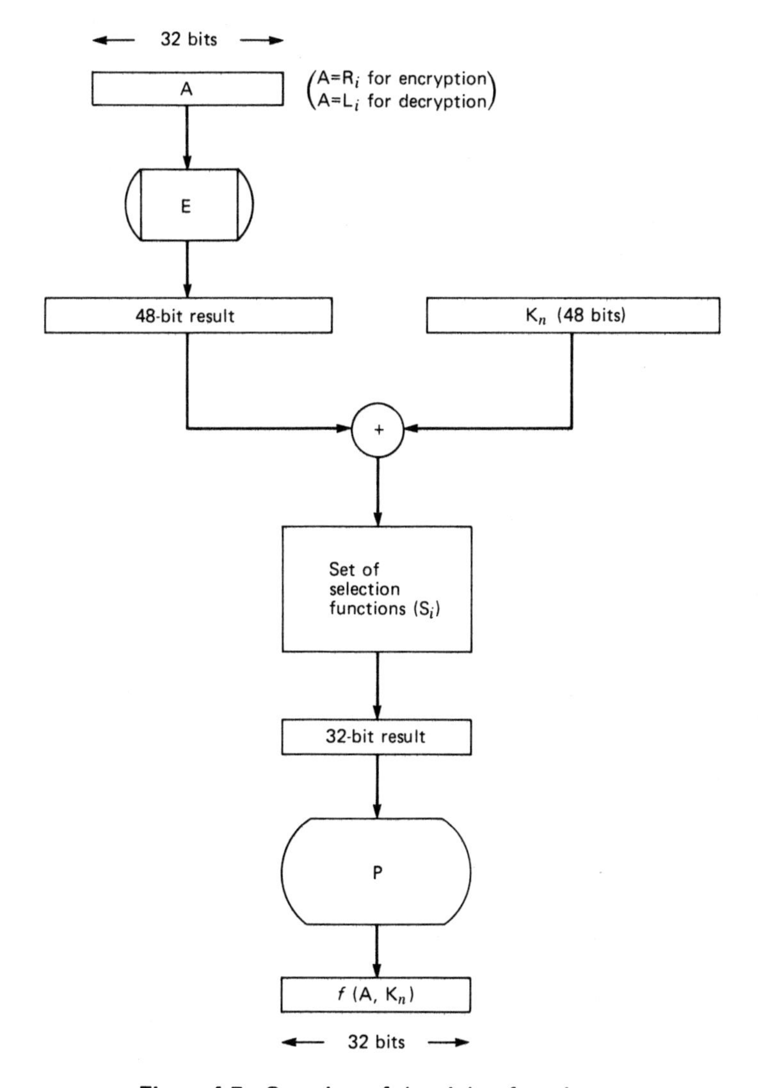

Figure 4.5 Overview of the cipher function

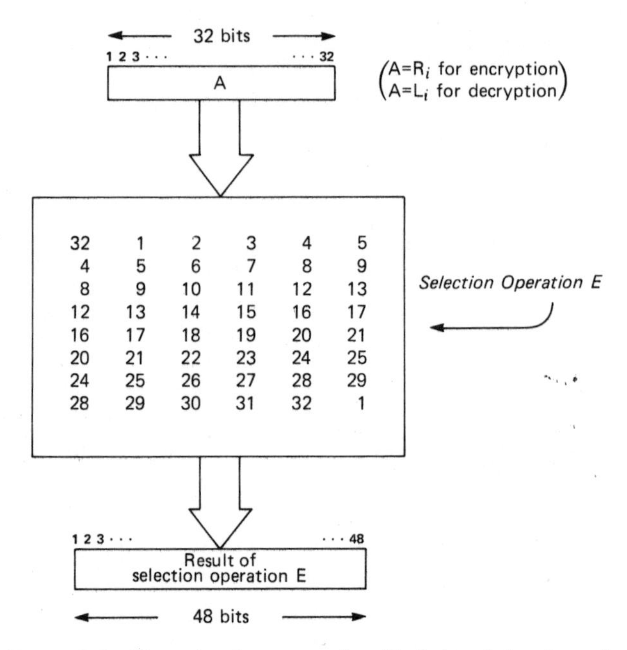

Figure 4.6 The selection operation *E* of the cipher function

The selection operation E, depicted in Figure 4.6, yields a 48-bit result wherein the bits of the result are respectively bits 32, 1, 2, . . . , 5, 4, 5, . . . , 9, 8, 9, . . . , 13, 12, 13, . . . , 17, 16, 17, . . . , 21, 20, 21, . . . , 25, 24, 25, . . . , 29, 28, 29, . . . , 1 of the symbolic argument A, which represents R_i or L_i, for encryption and decryption, respectively.

The unique set of selection functions S_i are shown conceptually in Figure 4.7, in which S_i takes a 6-bit block as input and yields a 4-bit result. A selection function is represented as a 4 X 16 matrix of numbers used in a prescribed manner.

Input to the unique set of selection functions S_1 through S_8 is a 48-bit block, denoted symbolically* as $B_1B_2B_3B_4B_5B_6B_7B_8$. Each B_i contains 8 bits. S_1 is to be used with B_1; S_2 is to be used with B_2; and so forth. If S_i is a selection function, and B_i is its argument, then the 4-bit result of the selection function is denoted by $S_i(B_i)$.

*$B_1B_2B_3B_4B_5B_6B_7B_8$ is intended to indicate the block consisting of the bits of B_1 followed by the bits of B_2 followed by the bits of B_3, . . . , followed by the bits of B_8.

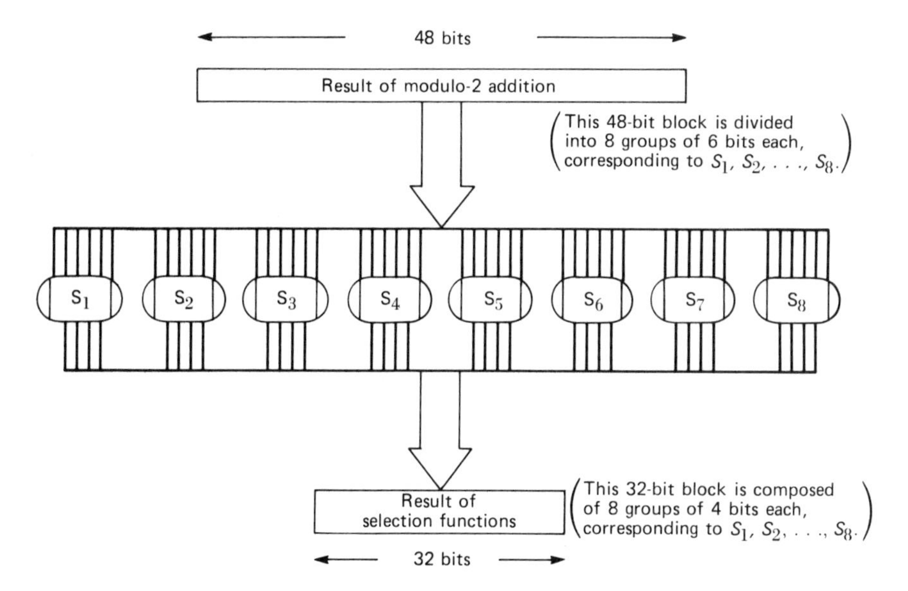

Figure 4.7 Overview of the unique set of selection functions used in the cipher function (note that each S_i takes 6 bits as input and yields 4 bits as output)

The result of the selection function S_i on argument B_i, denoted symbolically as $S_i(B_i)$, is computed as follows:

1. The first and last bits of B_i represent a binary number in the range zero to three, denoted by *m*.
2. The middle four bits of B_i represent a binary number in the range zero through fifteen, denoted by *n*.
3. Using zero-origin indexing, the number located in the mth row and n^{th} column of S_i's matrix is selected as a four-bit binary block.
4. The result of step 3 is the output of the selection function S_i.

The output of the complete set of selection functions, therefore, is the bit string $S_1(B_1)S_2(B_2)S_3(B_3)S_4(B_4)S_5(B_5)S_6(B_6)S_7(B_7)S_8(B_8)$ which denotes symbolically the 4-bit output of S_1, followed by the 4-bit output of S_2, and so forth.

Figure 4.8 contains an example of the use of selection function

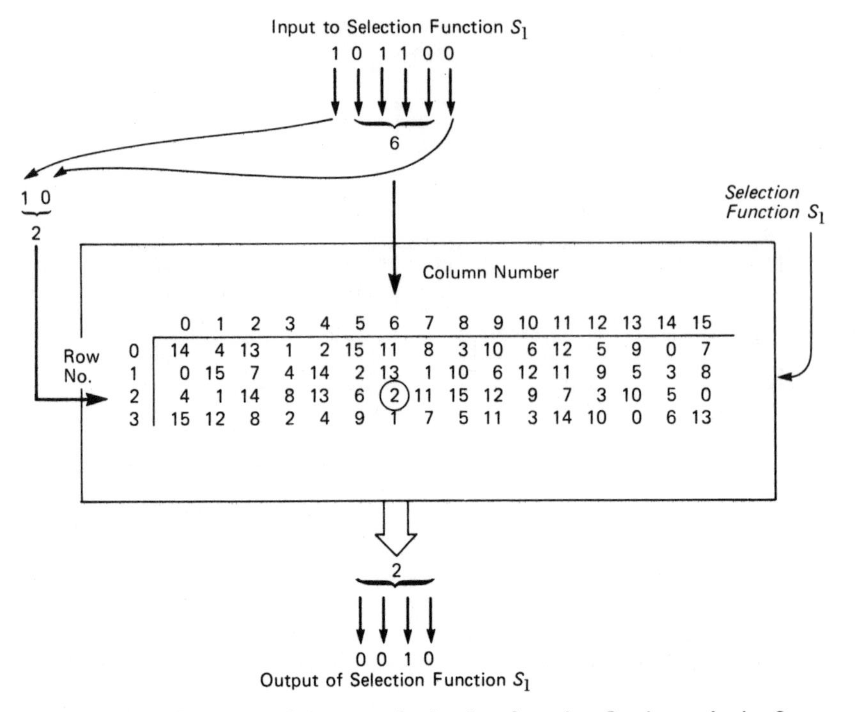

Figure 4.8 Example of the use of selection function S_1. Input is the 6-bit string 101100. Output is the 4-bit string 0010.

S_1. Input to the function is a following string of six bits: 101100. The first and last bits of the input are 1 and 0, respectively, which are the binary representation of 2. This is the row index. The middle four bits of the input are 0110, which is a binary representation of 6. This is the column index. Using 0-origin indexing, the value located in row 2 and column 6 of S_1's matrix is the value 2. The value of 2 is then converted to the four-bit binary value 0010, which is the four-bit output of selection function S_1 for the given input.

Table 4.2 gives the matrices corresponding to the selection functions S_1 through S_8.

The output of the set of eight selection functions, S_1 through S_8, is a string of 32 bits, as shown in Figure 4.7. This 32-bit output goes through a permutation operation *P* which yields a 32-bit result and completes the cipher function.* This final permutation in the

*Note that *P* does not complete the algorithm, but only the cipher function denoted symbolically by $f(A, K_n)$.

Table 4.2 Matrices for the selection functions S_1 through S_8

S_1

14	4	13	1	2	15	11	8	3	10	6	12	5	9	0	7
0	15	7	4	14	2	13	1	10	6	12	11	9	5	3	8
4	1	14	8	13	6	2	11	15	12	9	7	3	10	5	0
15	12	8	2	4	9	1	7	5	11	3	14	10	0	6	13

S_2

15	1	8	14	6	11	3	4	9	7	2	13	12	0	5	10
3	13	4	7	15	2	8	14	12	0	1	10	6	9	11	5
0	14	7	11	10	4	13	1	5	8	12	6	9	3	2	15
13	8	10	1	3	15	4	2	11	6	7	12	0	5	14	9

S_3

10	0	9	14	6	3	15	5	1	13	12	7	11	4	2	8
13	7	0	9	3	4	6	10	2	8	5	14	12	11	15	1
13	6	4	9	8	15	3	0	11	1	2	12	5	10	14	7
1	10	13	0	6	9	8	7	4	15	14	3	11	5	2	12

S_4

7	13	14	3	0	6	9	10	1	2	8	5	11	12	4	15
13	8	11	5	6	15	0	3	4	7	2	12	1	10	14	9
10	6	9	0	12	11	7	13	15	1	3	14	5	2	8	4
3	15	0	6	10	1	13	8	9	4	5	11	12	7	2	14

S_5

2	12	4	1	7	10	11	6	8	5	3	15	13	0	14	9
14	11	2	12	4	7	13	1	5	0	15	10	3	9	8	6
4	2	1	11	10	13	7	8	15	9	12	5	6	3	0	14
11	8	12	7	1	14	2	13	6	15	0	9	10	4	5	3

S_6

12	1	10	15	9	2	6	8	0	13	3	4	14	7	5	11
10	15	4	2	7	12	9	5	6	1	13	14	0	11	3	8
9	14	15	5	2	8	12	3	7	0	4	10	1	13	11	6
4	3	2	12	9	5	15	10	11	14	1	7	6	0	8	13

Table 4.2 Continued

S_7

4	11	2	14	15	0	8	13	3	12	9	7	5	10	6	1
13	0	11	7	4	9	1	10	14	3	5	12	2	15	8	6
1	4	11	13	12	3	7	14	10	15	6	8	0	5	9	2
6	11	13	8	1	4	10	7	9	5	0	15	14	2	3	12

S_8

13	2	8	4	6	15	11	1	10	9	3	14	5	0	12	7
1	15	13	8	10	3	7	4	12	5	6	11	0	14	9	2
7	11	4	1	9	12	14	2	0	6	10	13	15	3	5	8
2	1	14	7	4	10	8	13	15	12	9	0	3	5	6	11

cipher function is given in Figure 4.9. The permutation operation P yields a 32-bit result wherein the bits of the result are respectively, bits 16, 7, 20, 21, 29, . . . , 17, 1, . . . , 26, 5, . . . , 10, 2, . . . , 14, 32, . . . , 9, 19, . . . , 6, 22, . . . , 25 of the 32-bit result of the set of selection functions.

This completes the computations of the cipher function.

4.5 PREOUTPUT BLOCK

The output of the last iteration in the product transformation goes through a block transformation yielding a 64-bit result termed the *preoutput block.* The block transformation is the simple exchange of R_{16} and L_{16}, as depicted in Figure 4.10. The preoutput block is comprised of the bits of R_{16} followed by the bits of L_{16} and constitutes a 64-bit block, whose bits are numbered from 1 to 64 going from left to right.

4.6 INITIAL PERMUTATION

The initial permutation is the first step in the standard data encryption algorithm and is the key-independent permutation given in Figure 4.11. The output of the initial permutation are,

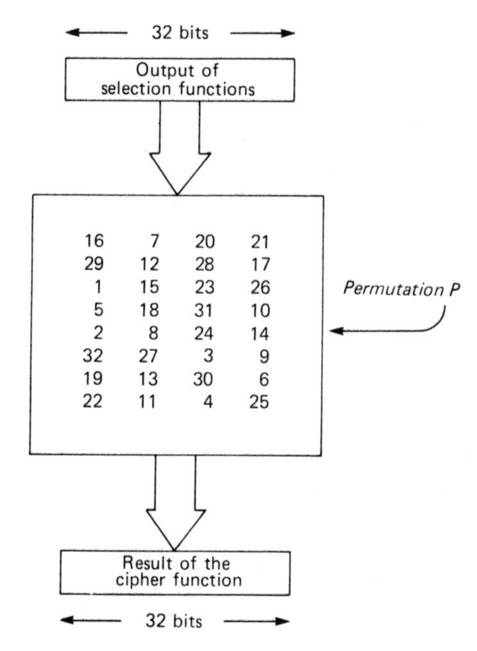

Figure 4.9 The permutation operation *P* of the cipher function

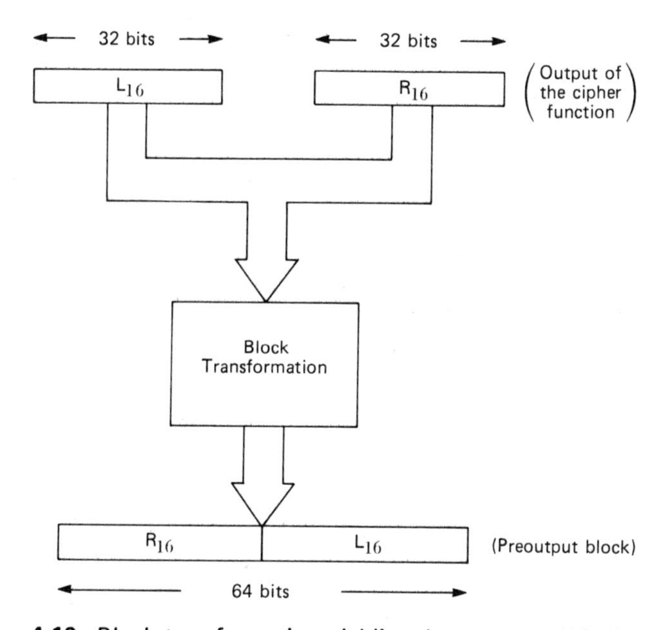

Figure 4.10 Block transformation yielding the *preoutput block*

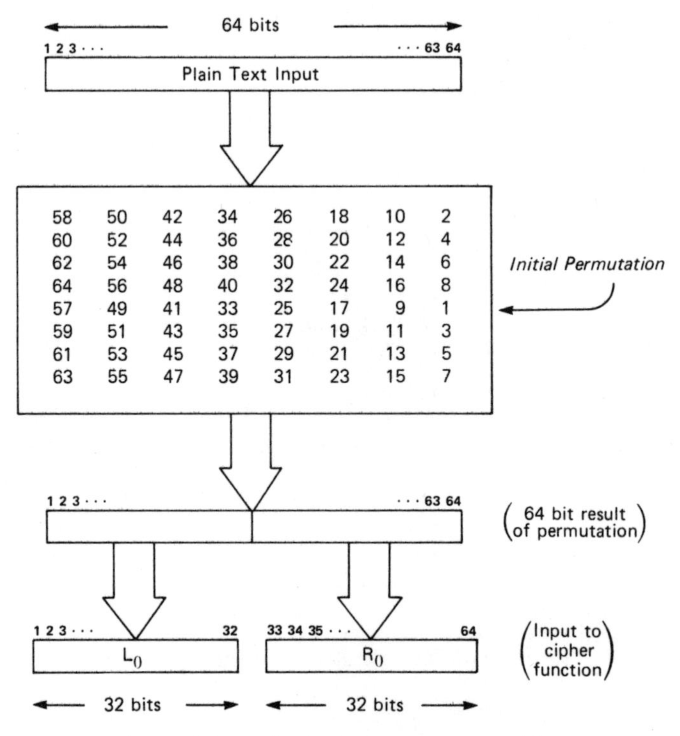

Figure 4.11 The initial permutation (IP)

respectively, bits 58, 50, . . . , 2, 60, . . . , 4, 62, . . . , . . . , 61, 53, . . . , 5, 63, . . . , 7 of the plain text input to the standard data encryption algorithm.

The result of the initial permutation (IP) is a 64-bit block. The leftmost 32 bits constitute L_0; the rightmost 32 bits constitute R_0. L_0 and R_0 are the initial input blocks to the product transformation.

4.7 INVERSE INITIAL PERMUTATION

The output of the product transformation is the preoutput block, which is subjected to a permutation which is the inverse of the initial permutation. The inverse initial permutation (IP^{-1}) is given in Figure 4.12. The output of IP^{-1}, which is synonymously

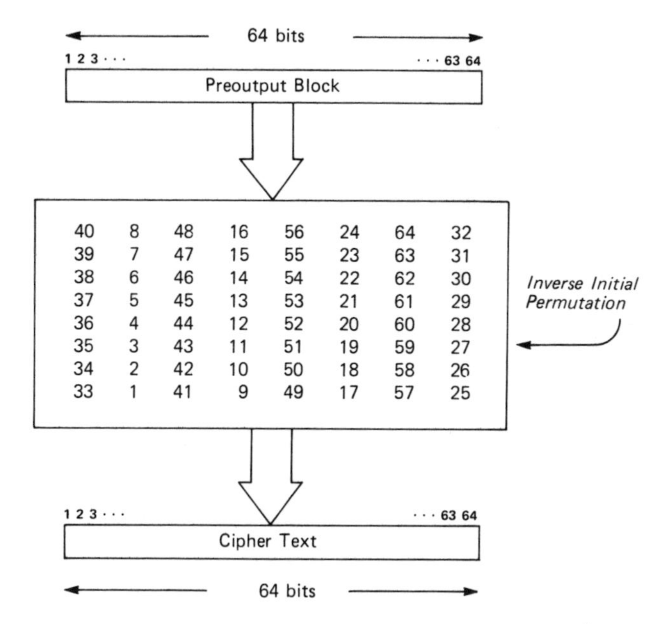

Figure 4.12 The inverse initial permutation (IP^{-1})

the cipher text output of the algorithm, is, respectively, bits 40, 8, . . . , 32, 39, . . . , 31, 38, . . . , . . . , 34, 2, . . . , 26, 33, 1, . . . , 25 of the preoutput block.

The 64-bit cipher text output of the standard data encryption algorithm can be used as a string of data bits for transmission or storage, or they may be converted back into BCD characters for subsequent data processing.

4.8 THE ENCIPHERING PROCESS

The enciphering process can be conveniently summarized symbolically. Given two blocks L and R and using the convention that LR denotes the block consisting of the bits of L followed by the bits of R, the initial permutation (IP) is specified as:

$$L_0R_0 \leftarrow IP(\langle \text{64-bit input block} \rangle)$$

Then if KS denotes the key schedule calculations, wherein the function KS yields a 48-bit subkey K_n for input arguments n and KEY, where KEY is the 64-bit cipher key, it follows that

$$K_n \leftarrow KS(n, KEY)$$

denotes the calculation of subkey K_n.

The 16 iterations in the product transformation that utilize the cipher function are then represented symbolically as:

$$L_n \leftarrow R_{n-1}$$
$$R_n \leftarrow L_{n-1} \oplus f(R_{n-1}, K_n)$$

where f is the cipher function and $\oplus$ denotes bit-by-bit modulo-2 addition. L_n and R_n are computed as n goes from 1 to 16. The pre-output block is $R_{16}L_{16}$ and the result of the algorithm is specified as:

$$\langle \text{64-bit cipher text} \rangle \leftarrow IP^{-1}(R_{16}L_{16})$$

The enciphering equations are summarized in Table 4.3.

Table 4.3 Summarization of the enciphering and deciphering equations

$$L_0R_0 \leftarrow IP(\langle \text{64-bit input block} \rangle)$$
$$L_n \leftarrow R_{n-1}$$
$$R_n \leftarrow L_{n-1} \oplus f(R_{n-1}, K_n)$$
$$\langle \text{64-bit cipher text} \rangle \leftarrow IP^{-1}(R_{16}L_{16})$$

Enciphering Equations

$$R_{16}L_{16} \leftarrow IP(\langle \text{64-bit cipher text} \rangle)$$
$$R_{n-1} \leftarrow L_n$$
$$L_{n-1} \leftarrow R_n \oplus f(L_n, K_n)$$
$$\langle \text{64-bit plain text} \rangle \leftarrow IP^{-1}(L_1R_1)$$

Deciphering Equations

4.9 THE DECIPHERING PROCESS

The process of deciphering a 64-bit cipher text block involves the same algorithm as encipherment, as stated in FIPS publication 46:[1]

> . . . to decipher it is only necessary to apply the very same algorithm to an enciphered message block, taking care that at each iteration of the computation the same block of key bits K is used during decipherment as was used during the encipherment of the block.

This is precisely the case because the initial permutation and the inverse initial permutation are by definition inverses of each other.

Applying the notation given previously, the result of the initial permutation (IP) is given as:

$$R_{16}L_{16} \leftarrow IP(\langle\text{64-bit cipher text}\rangle)$$

where the expression takes the final block transformation into consideration. The 16 iterations in the product transformation are represented symbolically as:

$$R_{n-1} \leftarrow L_n$$
$$L_{n-1} \leftarrow R_n \oplus f(L_n, K_n)$$

where the L_n and R_n are computed as n goes from 16 to 1. The result of the decipherment is then specified as:

$$\langle\text{64-bit plain text}\rangle \leftarrow IP^{-1}(L_0R_0)$$

The deciphering equations are also summarized in Table 4.3.

Appendix A contains a summary diagram of the enciphering computation.

---

[1]*Data Encryption Standard,* U.S. Department of Commerce, National Bureau of Standards, FIPS publication 46, 1977 January 15, p. 10.

5 BITWISE WALK-THROUGH OF THE STANDARD DATA ENCRYPTION ALGORITHM

5.1 INTRODUCTION

This chapter gives a sample case study of the use of the standard data encryption algorithm, showing all intermediate values in bitwise form. The objective is to clarify the methodology and, possibly, to implicitly answer questions concerning the algorithm that may have arisen in the reading of the previous chapter.

The APL functions used to generate the bit patterns displayed in this chapter are listed in Appendix C. The functions in Appendix C reflect modifications to the functions that comprise the formal definition of the standard data encryption algorithm to allow output to be printed at each stage of the computations so that a bitwise walk-through could be generated. The formal definition of the standard data encryption algorithm is given in chapter 6.

5.2 SAMPLE CASE STUDY

The plain text that is enciphered (i.e., encrypted) in this bitwise walk-through is the character string "RETRIEVE", where the quote marks are not part of the text and serve as delimiters. The cipher key is "FEBRUARY", where as before, the quote marks are not part of the character string. The plain text and the cipher key

```
        Q
58 50 42 34 26 18 10 2 60 52 44 36 28 20 12 4 62 54
      46 38 30 22 14 6 64 56 48 40 32 24 16 8 57 49
      41 33 25 17 9 1 59 51 43 35 27 19 11 3 61 53 45
      37 29 21 13 5 63 55 47 39 31 23 15 7

        E
32 1 2 3 4 5 4 5 6 7 8 9 8 9 10 11 12 13 12 13 14 15
      16 17 16 17 18 19 20 21 20 21 22 23 24 25 24 25
      26 27 28 29 28 29 30 31 32 1

        P
16 7 20 21 29 12 28 17 1 15 23 26 5 18 31 10 2 8 24
      14 32 27 3 9 19 13 30 6 22 11 4 25

        SHFT
1 1 2 2 2 2 2 2 1 2 2 2 2 2 2 1

        PC1A
57 49 41 33 25 17 9 1 58 50 42 34 26 18 10 2 59 51 43
      35 27 19 11 3 60 52 44 36

        PC1B
63 55 47 39 31 23 15 7 62 54 46 38 30 22 14 6 61 53
      45 37 29 21 13 5 28 20 12 4

        PC2
14 17 11 24 1 5 3 28 15 6 21 10 23 19 12 4 26 8 16 7
      27 20 13 2 41 52 31 37 47 55 30 40 51 45 33 48
      44 49 39 56 34 53 46 42 50 36 29 32
```

Figure 5.1 Printout of the permutation and selection matrices

both have a length of 8 characters that are represented as 8-bit bytes.

5.3 INITIALIZATION

The first step in enciphering (i.e., encryption) and deciphering (i.e., decryption) is to initialize the permutation and selection matrices. These matrices are listed here for completeness and are precisely the ones given in chapter 4. The permutation and selection matrices are given in Figure 5.1. The following identification is used for the matrices in Figure 5.1:

```
        S
14  4 13  1  2 15 11  8  3 10  6 12  5  9  0  7
 0 15  7  4 14  2 13  1 10  6 12 11  9  5  3  8
 4  1 14  8 13  6  2 11 15 12  9  7  3 10  5  0
15 12  8  2  4  9  1  7  4 11  1 14 10  0  6 13

15  1  8 14  6 11  3  4  9  7  1 13 12  0  5 10
 3 13  4  7 15  2  8 14 12  0  1 10  6  9 11  5
 0 14  7 11 10  4 13  1  5  8 12  6  9  3  2 15
13  8 10  1  3 15  4  2 11  6  7 12  0  5 14  9

10  0  9 14  6  3 15  5  1 13 12  7 11  4  2  8
13  7  0  9  3  4  6 10  2  8  5 14 12 11 15  1
13  6  4  9  8 15  3  0 11  1  2 12  5 10 14  7
 1 10 13  0  6  9  8  7  4 15 14  3 11  5  2 12

 7 13 14  3  0  6  9 10  1  2  8  5 11 12  4 15
13  8 11  5  6 15  0  3  4  7  2 12  1 10 14  9
10  6  9  0 12 11  7 13 15  1  3 14  5  2  8  4
 3 15  0  6 10  1 13  8  9  4  5 11 12  7  2 14

 2 12  4  1  7 10 11  6  8  5  3 15 13  0 14  9
14 11  2 12  4  7 13  1  5  0 15 10  3  9  8  6
 4  2  1 11 10 13  7  8 15  9 12  5  6  3  0 14
11  8 12  7  1 14  2 13  6 15  0  9 10  4  5  3

12  1 10 15  9  2  6  8  0 13  3  4 14  7  5 11
10 15  4  2  7 12  9  5  6  1 13 14  0 11  3  8
 9 14 15  5  2  8 12  3  7  0  4 10  1 13 11  6
 4  3  2 12  9  5 15 10 11 14  1  7  6  0  8 13

 4 11  2 14 15  0  8 13  3 12  9  7  5 10  6  1
13  0 11  7  4  9  1 10 14  3  5 12  2 15  8  6
 1  4 11 13 12  3  7 14 10 15  6  8  0  5  9  2
 6 11 13  8  1  4 10  7  9  5  0 15 14  2  3 12

13  2  8  4  6 15 11  1 10  9  3 14  5  0 12  7
 1 15 13  8 10  3  7  4 12  5  6 11  0 14  9  2
 7 11  4  1  9 12 14  2  0  6 10 13 15  3  5  8
 2  1 14  7  4 10  8 13 15 12  9  0  3  5  6 11
```

Figure 5.1 Continued

Matrix	*Permutation*
Q	Initial permutation
E	Used in the cipher function
SHFT	Used to calculate the key schedule
P	Used in the cipher function
S	Selection functions
PC1A	Permuted choice 1–C_0
PC1B	Permuted choice 1–D_0
PC2	Permuted choice 2

The function for generating the permutation and selection matrices is listed in chapter 6.

5.4 COMPUTING THE KEY SCHEDULE

The first step in computing the key schedule is to encode* the characters of the cipher key as 8-bit bytes. Using the cipher key "FEBRUARY", the characters are encoded as follows:

Character	*Bit Pattern*
F	01011011
E	01011010
B	01010111
R	01100111
U	01101010
A	01010110
R	01100111
Y	01101110

The 8-bit bytes are then catenated as follows:

```
01011011 01011010 01010111 01100111 01101010 01010110 01100111 0
    1101110
```

to form the 64-bit cipher key:

```
0101101101011010010101110110011101101010010101100110011101101110
```

After going through permuted choice 1, where C_0 and D_0 are generated, the sixteen iterations necessary to generate the key schedule and the 16 subkeys are:

```
N =  1  C = 0000000111111111101100000100  D = 11111111110110010
        01001101111
SUBKEY = 1110000001001011011100110101111111010100111001111
```

```
N =  2  C = 0000001111111111011000001000  D = 11111111101100100
        10011011111
SUBKEY = 101000000100101100111001001110111001001111101111
```

*The bit patterns used for the characters are the 8-bit encoded values of the index of that character in the APL atomic vector. This information is contained in Appendix B.

```
N =  3  C = 0000111111111110110000001000000  D = 11111110110010010
      01101111111
SUBKEY = 111001000101101001110010111111101011100111001011

N =  4  C = 0011111111110110000001000000000  D = 11111011001001001
      10111111111
SUBKEY = 101001101111001101010000011001101111011101111111

N =  5  C = 1111111111011000000100000000000  D = 11101100100100110
      11111111111
SUBKEY = 000011100101011101010011011111111011110111101010

N =  6  C = 1111111011000000100000000000011  D = 10110010010011011
      11111111111
SUBKEY = 011011110101000101011001111011001101110101111011

N =  7  C = 1111110110000001000000000001111  D = 11001001001101111
      11111111110
SUBKEY = 000011111100000111001001010011111111111001111110

N =  8  C = 1111011000000100000000000111111  D = 00100100110111111
      11111111011
SUBKEY = 000110110100100110011011111111011101110111111000

N =  9  C = 1110110000001000000000001111111  D = 01001001101111111
      11111110110
SUBKEY = 000111110100101010001001110111111111111000111100

N = 10  C = 1011000000100000000000111111111  D = 00100110111111111
      11111011001
SUBKEY = 000110110011100110001100111110010101111111111000

N = 11  C = 1100000100000000000111111111110  D = 10011011111111111
      11101100100
SUBKEY = 000110000010110011001101100110011111101000111111

N = 12  C = 0000010000000000011111111111011  D = 01101111111111111
      10110010010
SUBKEY = 010100010110110000101100111011101111110101101100

N = 13  C = 0001000000000001111111111101100  D = 10111111111111110
      11001001001
SUBKEY = 110000001010110110100100101110010010101111111111

N = 14  C = 0100000000001111111111110110000  D = 11111111111111011
      00100100110
SUBKEY = 110100001010111000100111011011111111010100101111

N = 15  C = 0000000001111111111011000001  D = 11111111111101100
      10010011011
SUBKEY = 111000011011011000100010011101110010011111110111

N = 16  C = 0000000011111111110110000010  D = 11111111111011001
      00100110111
SUBKEY = 111000001011001000101110111101101111001110010111
```

For each iteration, C_i and D_i are circularly left shifted the specified number of places, and then C_i and D_i are then tapped off and subjected to permuted choice 2 to establish the ith subkey. Each of the computations is reflected in the above listing. The required shift amounts are given in an earlier section entitled "Initialization."

5.5 THE ENCRYPTION PROCESS

The steps in the process of encryption are covered sequentially as they take place in real time. The output of each computation is interspersed in the text.

Encoding the Plain Text

The first step in encryption is to encode the characters of the plain text as 8-bit bytes. Using the plain text "RETRIEVE", the characters are encoded as follows:

Character	*Bit Pattern*
R	01100111
E	01011010
T	01101001
R	01100111
I	01011110
E	01011010
V	01101011
E	01011010

The 8-bit bytes are then catenated as follows:

```
01100111 01011010 01101001 01100111 01011110 01011010 01101011 0
     1011010
```

to form the 64-bit plain text, referred to as the input block:

```
0110011101011010011010010110011101011110010110100110101101011010
```

Permuted Input Block

The input block is first subjected to the initial permutation, given as follows:

```
INPUT BLOCK = 01100111010110100110100101100111010111100101101001
      10101101011010
PERMUTTED INPUT BLOCK = 1111111110110010000110010100110100000000
      01001101111110110111111011
```

The leftmost 32 bits of the permuted input block become L_0 and the rightmost 32 bits of the permuted input block become R_0:

```
L(ZERO) = 11111111101100100001100101001101   R(ZERO) = 0000000001
      0011011111011011111011
```

L_0 and R_0 serve as the starting data for the complex product transformation. In the product transformation, the 16 iterations of the cipher function are referred to as the "Forward Cipher Function," to distinguish it from the decryption process.

The Forward Cipher Function

The forward cipher function is comprised of 16 iterations. The input to the ith iteration is L_{i-1}, R_{i-1}, and K_i; the output of each iteration is L_i and R_i. The computed results from each computational step in the 16 iterations is given as follows:

```
CIPHER ITERATION:  1
CIPHER:  L = 11111111101100100001100101001101   R = 0000000001001
      1011111011011111011
KEY SCH:  K = 111000001001011011100110101111111010100111001111
F:  E[R] = 100000000000001001011011111110101101011111110110
F:  E[R]⊕K = 011000001001010010111101010001010111111000111001
F:  OUTP OF SEL FCN = 01011111110100100101111000000011
F:  OUTP OF PERM P = 01110100011011111100100100011010
CIPHER:  F(R,K) = 01110100011011111100100100011010
CIPHER:  L⊕F(R,K) = 10001011110111011101000001010111
CIPHER: NEXT L = 00000000010011011111011011111011   NEXT R = 1000
      1011110111011101000001010111


CIPHER ITERATION:  2
CIPHER:  L = 00000000010011011111011011111011   R = 100010111011
      10111010000010101111
KEY SCH:  K = 101000001001011001110010011101110010011111101111
F:  E[R] = 110001010111111011111101111101010000000010101011111
F:  E[R]⊕K = 011001011110100010001001100111010010010101000000
F:  OUTP OF SEL FCN = 10011010011001100111101010111101
F:  OUTP OF PERM P = 01111010110111010011100010101110
CIPHER:  F(R,K) = 01111010110111010011100010101110
CIPHER:  L⊕F(R,K) = 01111010100100001100111001010101
CIPHER: NEXT L = 10001011110111011101000001010111   NEXT R = 0111
      1010100100001100111001010101
```

```
CIPHER ITERATION:  3
CIPHER:  L = 10001011110111011101000001010111  R = 0111101010010
      0001100111001010101
KEY SCH:  K = 111001000101101001110010111111101011100111001011
F:  E[R] = 101111110101010010100001011001011100001010101010
F:  E[R]⊕K = 010110110000111011010011100110110111101101100001
F:  OUTP OF SEL FCN = 11000101010101111011011110100010
F:  OUTP OF PERM P = 10100101111000111111010010011001
CIPHER:  F(R,K) = 10100101111000111111010010011001
CIPHER:  L⊕F(R,K) = 00101110001111100010010011001110
CIPHER: NEXT L = 01111010100100001100111001010101  NEXT R = 0010
      1110001111100010010011001110


CIPHER ITERATION:  4
CIPHER:  L = 01111010100100001100111001010101  R = 0010111000111
      1100010010011001110
KEY SCH:  K = 101001101111001101010000011001101111011101111111
F:  E[R] = 000101011100000111111100000010000100101100101100
F:  E[R]⊕K = 101100110011001010101100011101100110000100100011
F:  OUTP OF SEL FCN = 00100110001101111000010100100001
F:  OUTP OF PERM P = 11000101010000000011111000011100
CIPHER:  F(R,K) = 11000101010000000011111000011100
CIPHER:  L⊕F(R,K) = 10111111110100001111000001001001
CIPHER: NEXT L = 00101110001111100010010011001110  NEXT R = 1011
      1111110100001111000001001001


CIPHER ITERATION:  5
CIPHER:  L = 00101110001111100010010011001110  R = 1011111111010
      0001111000001001001
KEY SCH:  K = 000011100101011101010011011111111011110111101010
F:  E[R] = 110111111111111010100001011110100000001001010011
F:  E[R]⊕K = 110100011010100111110010000001011011111110111001
F:  OUTP OF SEL FCN = 10010000000000011110101100100011
F:  OUTP OF PERM P = 10010001101001100010110010000010
CIPHER:  F(R,K) = 10010001101001100010110010000010
CIPHER:  L⊕F(R,K) = 10111111100110000000100001001100
CIPHER: NEXT L = 10111111110100001111000001001001  NEXT R = 1011
      1111100110000000100001001100


CIPHER ITERATION:  6
CIPHER:  L = 10111111110100001111000001001001  R = 1011111110011
      0000000100001001100
KEY SCH:  K = 011011110101000101011001111011001101110101111011
F:  E[R] = 010111111111110011110000000001010000001001011001
F:  E[R]⊕K = 001100001010110110101001111010011101111100100010
F:  OUTP OF SEL FCN = 10111011110010100011001110011011
F:  OUTP OF PERM P = 01101010111010110110101111000011
CIPHER:  F(R,K) = 01101010111010110110101111000011
CIPHER:  L⊕F(R,K) = 11010101001110111001101110001010
CIPHER: NEXT L = 10111111100110000000100001001100  NEXT R = 1101
      0101001110111001101110001010
```

```
CIPHER ITERATION:  7
CIPHER:  L = 10111111100110000000100001001100   R = 1101010100111
      0111001101110001010
KEY SCH:  K = 000011111100000111001001010011111111111001111110
F:  E[R] = 011010101010100111110111110011110111110001010101
F:  E[R]⊕K = 011001010110100000111110100000000100000100010101l
F:  OUTP OF SEL FCN = 10011101110101000100100111111010
F:  OUTP OF PERM P = 00011110100111110111010100010011
CIPHER:  F(R,K) = 00011110100111110111010100010011
CIPHER:  L⊕F(R,K) = 10100001000001110111110101011111
CIPHER: NEXT L = 11010101001110111001101110001010  NEXT R = 1010
      0001000001110111110101011111


CIPHER ITERATION:  8
CIPHER:  L = 11010101001110111001101110001010   R = 101000010000
      01110111110101011111
KEY SCH:  K = 000110110100100110011011111111011101110111111000
F:  E[R] = 110100000010100000001110101111111010101011111111
F:  E[R]⊕K = 110010110110000110010101010000100111011100000111
F:  OUTP OF SEL FCN = 11000110111000101000110001101000
F:  OUTP OF PERM P = 01011001110100011000010100011100
CIPHER:  F(R,K) = 01011001110100011000010100011100
CIPHER:  L⊕F(R,K) = 10001100111010100001111010010110
CIPHER: NEXT L = 10100001000001110111110101011111  NEXT R = 1000
      1100111010100001111010010110


CIPHER ITERATION:  9
CIPHER:  L = 10100001000001110111110101011111   R = 100011001110
      10100001111010010110
KEY SCH:  K = 000111110100101010001001110111111111111000111100
F:  E[R] = 010001011001011101010100000011111101010010101101
F:  E[R]⊕K = 010110101101110111011101110100000010101010010001
F:  OUTP OF SEL FCN = 11000100001111011000000100111100
F:  OUTP OF PERM P = 00001111110001001011010001110100
CIPHER:  F(R,K) = 00001111110001001011010001110100
CIPHER:  L⊕F(R,K) = 10101110110000111100100100101011
CIPHER: NEXT L = 10001100111010100001111010010110  NEXT R = 1010
      1110110000111100100100101011


CIPHER ITERATION: 10
CIPHER:  L = 10001100111010100001111010010110   R = 101011101100
      00111100100100101011
KEY SCH:  K = 000110110011100110001100111110010101111111111000
F:  E[R] = 110101011101011000000111111001010010100101010111
F:  E[R]⊕K = 110011101110111110001011000111000111011010101111
F:  OUTP OF SEL FCN = 10110001011111111110000101010110l
F:  OUTP OF PERM P = 10001101111001010101111001100111
CIPHER:  F(R,K) = 10001101111001010101111001100111
CIPHER:  L⊕F(R,K) = 00000001000011110100000001111000l
CIPHER: NEXT L = 10101110110000111100100100101011  NEXT R = 0000
      0001000011110100000001111000l
```

```
CIPHER ITERATION: 11
CIPHER:  L = 10101110110000111100100100101011  R = 0000000100001
      1110100000011110001
KEY SCH:  K = 000110000010110011001101100110011111101000111111
F:  E[R] = 100000000010100001011110101000000001011110100010
F:  E[R]⊕K = 100110000000010010010011001110011110110110011101
F:  OUTP OF SEL FCN = 10001111110101110110101110001001
F:  OUTP OF PERM P = 11011100111011010111100110010001
CIPHER:  F(R,K) = 11011100111011010111100110010001
CIPHER:  L⊕F(R,K) = 01110010001011101011000010111010
CIPHER: NEXT L = 00000001000011110100000011110001  NEXT R = 0111
      0010001011101011000010111010


CIPHER ITERATION: 12
CIPHER:  L = 00000001000011110100000011110001  R = 011100100010
      11101011000010111010
KEY SCH:  K = 010100010110110000101100111101110111111010110100
F:  E[R] = 001110100100000101011101010110100001010111110100
F:  E[R]⊕K = 011010110010110101110001101011010110101101000000
F:  OUTP OF SEL FCN = 10011000111010011110010010101101
F:  OUTP OF PERM P = 10001001100011010000110111101111
CIPHER:  F(R,K) = 10001001100011010000110111101111
CIPHER:  L⊕F(R,K) = 10001000100000100100110100011110
CIPHER: NEXT L = 01110010001011101011000010111010  NEXT R = 1000
      1000100000100100110100011110


CIPHER ITERATION: 13
CIPHER:  L = 01110010001011101011000010111010  R = 1000100010000
      0100100110100011110
KEY SCH:  K = 110000001010110110100100101110010010101111111111
F:  E[R] = 010001010001010000000100001001011010100011111101
F:  E[R]⊕K = 100001011011100110100000100111001000001100000010
F:  OUTP OF SEL FCN = 11111001100110100111100110000010
F:  OUTP OF PERM P = 00110100110011101110001111000011
CIPHER:  F(R,K) = 00110100110011101110001111000011
CIPHER:  L⊕F(R,K) = 01000110111000000101001101111001
CIPHER: NEXT L = 10001000100000100100110100011110  NEXT R = 0100
      0110111000000101001101111001


CIPHER ITERATION: 14
CIPHER:  L = 10001000100000100100110100011110  R = 010001101110
      00000101001101111001
KEY SCH:  K = 110100001010111000100111101101111111101010010111
F:  E[R] = 101000001101011100000000001010100110101111110010
F:  E[R]⊕K = 011100000111100100100111100111011001000110100101
F:  OUTP OF SEL FCN = 00000111010001100111000010111110
F:  OUTP OF PERM P = 01101010010001110101010010110001
CIPHER:  F(R,K) = 01101010010001110101010010110001
CIPHER:  L⊕F(R,K) = 11100010110001010001100110101111
CIPHER: NEXT L = 01000110111000000101001101111001  NEXT R = 1110
      0010110001010001100110101111
```

```
CIPHER ITERATION: 15
CIPHER:  L = 01000110111000000101001101111001  R = 1110001011000
      1010001100110101111
KEY SCH:  K = 111000011011011000100010011101110010011111110111
F:  E[R] = 111100000101011000001010100011110011110101011111
F:  E[R]⊕K = 000100011110000000101000111110000001101010101000
F:  OUTP OF SEL FCN = 11011010101011001110101000111001
F:  OUTP OF PERM P = 01011011101011001001110111000110
CIPHER:  F(R,K) = 01011011101011001001110111000110
CIPHER:  L⊕F(R,K) = 00011101010011001100111010111111
CIPHER: NEXT L = 11100010110001010001100110101111  NEXT R = 0001
      1101010011001100111010111111


CIPHER ITERATION: 16
CIPHER:  L = 11100010110001010001100110101111  R = 0001110101001
      1001100111010111111
KEY SCH:  K = 111000001011001000101110111101101111001110010111
F:  E[R] = 100011111010101001011001011001011101010111111110
F:  E[R]⊕K = 011011110001100001110111100100110010011001101001
F:  OUTP OF SEL FCN = 01011011000110110001000000100100
F:  OUTP OF PERM P = 11100100010010001100010001100010
CIPHER:  F(R,K) = 11100100010010001100010001100010
CIPHER:  L⊕F(R,K) = 00000110100011011101110111001101
CIPHER: NEXT L = 00011101010011001100111010111111  NEXT R = 0000
      0110100011011101110111001101
```

The final result of the 16th iteration is a set of two 32-bit blocks identified as L_{16} and R_{16}.

The Preoutput Block and the Inverse Initial Permutation

The preoutput block is computed as a block transformation of L_{16} and R_{16}. The inverse initial permutation is then applied to the preoutput block yielding the following computations:

```
PREOUTPUT BLOCK = 0000011010001101110111011100110100011101010011
      001100111010111111
PERMUTTED OUTPUT BLOCK = 100101110100101011111111101111111000011
      0000000100010110100011111
```

At this stage, the 64-bit cipher text is in binary form and would ordinarily be transmitted over a communications channel or be stored. In this case, it is converted back into character form so that it can be displayed.

Decoding the Cipher Text

The last step in the bitwise walk-through of the encryption process is to decode the 64 bits of the cipher text into character form. The 64-bit cipher text block:

```
1001011101001010111111111011111110000110000000100010110100011111
```

is first divided into 8-bit bytes as follows:

```
10010111 01001010 11111111 10111111 10000110 00000010 00101101 0
    0011111
```

and is then decoded into cipher text characters:

Bit Pattern	*Character*
10010111	¯
01001010	⌿
11111111	⍒
10111111	⍒
10000110	V
00000010	⍒̈
00101101	ρ
00011111	┌

The resulting cipher text in character form is:

¯⌿⍒⍒V⍒ρ┌

5.6 THE DECRYPTION PROCESS

The presentation of the process of decryption follows the pattern established for encryption. The first step is to perform initialization and then to compute the key schedule as covered in sections 5.3 and 5.4, respectively.

Encoding the Cipher Text

Decryption is required when the cipher text is received over communications facilities or is retrieved from storage and is needed in its plain text form for normal computer processing. In this case study, the cipher text exists in character form as a result of encryption—covered previously. In actual practice, the cipher text may originate in binary form for decryption.

The first step in this bitwise walk-through of decryption is to encode the characters of the cipher text as 8-bit bytes. Using the encrypted form of the plain text "RETRIEVE" with the cipher key "FEBRUARY" as input, the cipher text is encoded as follows:

Character	*Bit Pattern*
–	10010111
⁄	01001010
▯	11111111
▯	10111111
V	10000110
▯̈	00000010
ρ	00101101
⌈	00011111

The 8-bit bytes are then catenated as follows:

10010111 01001010 11111111 10111111 10000110 00000010 00101101 0
0011111

to form the 64-bit cipher text, referred to as the output block:

1001011101001010111111111011111110000110000000100010110100011111

It should be recognized in decryption that the encryption process is essentially reversed so the decryption process starts with the output block.

The Initial Permutation and the Preoutput Block

The output block is first subjected to the initial permutation, which reverses the inverse initial permutation applied during encryption, producing the following bit patterns:

```
OUTPUT BLOCK = 1001011101001010111111110111111100001100000000100
      010110100011111
PREOUTPUT BLOCK = 0000011010001101110111011100110100011101010011
      0011001110101111111
```

The block transformation employed during encryption is reversed so that the leftmost 32 bits of the preoutput block become R_{16} and the rightmost 32 bits of the preoutput block become L_{16}, as follows:

```
L(16) = 00011101010011001100111010111111   R(16) = 0000011010001 1
      0111011101110011101
```

L_{16} and R_{16} serve as the starting point for the sixteen iterations of the cipher function performed in reverse order as subkeys range from K_{16} to K_1. The sixteen iterations of the cipher function, in this case, are referred to as the "Reverse Cipher Function."

The Reverse Cipher Function

The reverse cipher function is comprised of 16 iterations. The input to the ith iteration is L_i, R_i, and K_i; the output of each iteration is L_{i-1} and R_{i-1}. The computed results from each computational step in the 16 iterations is given as follows:

```
DECIPHER ITERATION: 16
DECIPHER:  L = 00011101010011001100111010111111  R = 00000110100
      0110111011101110011101
KEY SCH:  K = 111000000101100100010111011110110111001110010111
F:  E[R] = 100011111010101001011001011001011101010111111110
F:  E[R]⊕K = 011011110001100001110111100100110010011001101001
F:  OUTP OF SEL FCN = 01011011000110110001000000100100
F:  OUTP OF PERM P = 11100100010010001100010001100010
DECIPHER:  F(L,K) = 11100100010010001100010001100010
DECIPHER:  R⊕F(L,K) = 11100010110001010001100110101111
DECIPHER:  PREV L = 11100010110001010001100110101111  PREV R = 0
      0011101010011001100111010111111
```

```
DECIPHER ITERATION: 15
DECIPHER:  L = 11100010110001010001100110101111  R = 00011101010
      011001100111010111111
KEY SCH:  K = 111000011011011000100010011101110010011111110111
F:  E[R] = 111100000101011000001010100011110011110101011111
F:  E[R]⊕K = 000100011110000000101000111110000001101010101000
F:  OUTP OF SEL FCN = 11011010101011001110101000111001
F:  OUTP OF PERM P = 01011011101011001001110111000110
DECIPHER:  F(L,K) = 01011011101011001001110111000110
DECIPHER:  R⊕F(L,K) = 01000110111000000101001101111001
DECIPHER:  PREV L = 01000110111000000101001101111001  PREV R = 1
      1100010110001010001100110101111


DECIPHER ITERATION: 14
DECIPHER:  L = 01000110111000000101001101111001  R = 11100010110
      001010001100110101111
KEY SCH:  K = 110100001010111000100111101101111111101010010111
F:  E[R] = 101000001101011100000000001010100110101111110010
F:  E[R]⊕K = 011100000111100100100111100111011001000101100101
F:  OUTP OF SEL FCN = 00000111010001100111000010111110
F:  OUTP OF PERM P = 01101010010001110101010010110001
DECIPHER:  F(L,K) = 01101010010001110101010010110001
DECIPHER:  R⊕F(L,K) = 10001000100000100100110100011110
DECIPHER:  PREV L = 10001000100000100100110100011110  PREV R = 0
      1000110111000000101001101111001


DECIPHER ITERATION: 13
DECIPHER:  L = 10001000100000100100110100011110  R = 01000110111
      000000101001101111001
KEY SCH:  K = 110000001010110110100100101110010010101111111111
F:  E[R] = 010001010001010000000100001001011010100011111101
F:  E[R]⊕K = 100001011011100110100000100111001000001100000010
F:  OUTP OF SEL FCN = 11111001100110100111100110000010
F:  OUTP OF PERM P = 00110100110011101110001111000011
DECIPHER:  F(L,K) = 00110100110011101110001111000011
DECIPHER:  R⊕F(L,K) = 01110010001011101011000010111010
DECIPHER:  PREV L = 01110010001011101011000010111010  PREV R = 1
      0001000100000100100110100011110


DECIPHER ITERATION: 12
DECIPHER:  L = 01110010001011101011000010111010  R = 1000100010
      0000100100110100011110
KEY SCH:  K = 010100010110110000101100111101110111111010110100
F:  E[R] = 001110100100000101011101010110100001010111110100
F:  E[R]⊕K = 011010110010110101110001101011010110101101000000
F:  OUTP OF SEL FCN = 10011000111010011110010010101101
F:  OUTP OF PERM P = 10001001100011010000110111101111
DECIPHER:  F(L,K) = 10001001100011010000110111101111
DECIPHER:  R⊕F(L,K) = 00000001000011110100000011110001
DECIPHER:  PREV L = 00000001000011110100000011110001  PREV R = 0
      1110010001011101011000010111010
```

```
DECIPHER ITERATION: 11
DECIPHER:  L = 00000001000011110100000011110001  R = 01110010001
      0111010110000101111010
KEY SCH:  K = 000110000010110011001101100110011111101000111111
F:  E[R] = 100000000010100001011110101000000001011110100010
F:  E[R]⊕K = 100110000000010010010011001110011110110110011101
F:  OUTP OF SEL FCN = 10001111110101110110101110001001
F:  OUTP OF PERM P = 11011100111011010111100110010001
DECIPHER:  F(L,K) = 11011100111011010111100110010001
DECIPHER:  R⊕F(L,K) = 10101110110000111100100100101011
DECIPHER:  PREV L = 10101110110000111100100100101011  PREV R = 0
      0000001000011110100000011110001


DECIPHER ITERATION: 10
DECIPHER:  L = 10101110110000111100100100101011  R = 00000001000
      011110100000011110001
KEY SCH:  K = 000110110011100110001100111110010101111111111000
F:  E[R] = 110101011101011000000111111001010010100101010111
F:  E[R]⊕K = 110011101110111110001011000111000111011010101111
F:  OUTP OF SEL FCN = 10110001011111111100001010101101
F:  OUTP OF PERM P = 10001101111001010101111001100111
DECIPHER:  F(L,K) = 10001101111001010101111001100111
DECIPHER:  R⊕F(L,K) = 10001100111010100001111010010110
DECIPHER:  PREV L = 10001100111010100001111010010110  PREV R = 1
      0101110110000111100100100101011


DECIPHER ITERATION:  9
DECIPHER:  L = 10001100111010100001111010010110  R = 10101110110
      000111100100100101011
KEY SCH:  K = 000111110100101010001001110111111111111000111100
F:  E[R] = 010001011001011101010100000011111101010010101101
F:  E[R]⊕K = 010110101101110111011101110100000010101010010001
F:  OUTP OF SEL FCN = 11000100001111101100000100111100
F:  OUTP OF PERM P = 00001111110001001011010001110100
DECIPHER:  F(L,K) = 00001111110001001011010001110100
DECIPHER:  R⊕F(L,K) = 10100001000001110111110101011111
DECIPHER:  PREV L = 10100001000001110111110101011111  PREV R = 1
      0001100111010100001111010010110


DECIPHER ITERATION:  8
DECIPHER:  L = 10100001000001110111110101011111  R = 10001100111
      010100001111010010110
KEY SCH:  K = 000110110100100110011011111111011101110111111000
F:  E[R] = 110100000010100000001110101111111010101011111111
F:  E[R]⊕K = 110010110110000110010101010000100111011100000111
F:  OUTP OF SEL FCN = 11000110111000101000110001101000
F:  OUTP OF PERM P = 01011001110100011000010100011100
DECIPHER:  F(L,K) = 01011001110100011000010100011100
DECIPHER:  R⊕F(L,K) = 11010101001110111001101110001010
DECIPHER:  PREV L = 11010101001110111001101110001010  PREV R = 1
      0100001000001110111110101011111
```

```
DECIPHER ITERATION:  7
DECIPHER:  L = 11010101001110111001101110001010   R = 10100001000
       001110111110101011111
KEY SCH:  K = 000011111100000111001001010011111111111001111110
F:  E[R] = 011010101010100111110111110011110111110001010101
F:  E[R]⊕K = 011001010110100000111110100000001000001000101011
F:  OUTP OF SEL FCN = 10011101110101000100100111111010
F:  OUTP OF PERM P = 00011110100111110111010100010011
DECIPHER:  F(L,K) = 00011110100111110111010100010011
DECIPHER:  R⊕F(L,K) = 10111111100110000000100001001100
DECIPHER:  PREV L = 10111111100110000000100001001100   PREV R = 1
       1010101001110111001101110001010

DECIPHER ITERATION:  6
DECIPHER:  L = 10111111100110000000100001001100   R = 11010101001
       110111001101110001010
KEY SCH:  K = 011011110101000101011001111011001101110101111011
F:  E[R] = 010111111111110011110000000001010000001001011001
F:  E[R]⊕K = 001100001010110110101001111010011101111100100010
F:  OUTP OF SEL FCN = 10111011110010100011001110011011
F:  OUTP OF PERM P = 01101010111010110110101111000011
DECIPHER:  F(L,K) = 01101010111010110110101111000011
DECIPHER:  R⊕F(L,K) = 10111111110100001111000001001001
DECIPHER:  PREV L = 10111111110100001111000001001001   PREV R = 1
       0111111100110000000100001001100

DECIPHER ITERATION:  5
DECIPHER:  L = 10111111110100001111000001001001   R = 10111111100
       110000000100001001100
KEY SCH:  K = 000011100101011101010011011111111101110111101010
F:  E[R] = 110111111111111010100001011110100000001001010011
F:  E[R]⊕K = 110100011010100111110010000001011101111110111001
F:  OUTP OF SEL FCN = 10010000000000011110101100100011
F:  OUTP OF PERM P = 10010001101001100010110010000010
DECIPHER:  F(L,K) = 10010001101001100010110010000010
DECIPHER:  R⊕F(L,K) = 00101110001111100010010011001110
DECIPHER:  PREV L = 00101110001111100010010011001110   PREV R = 1
       0111111110100001111000001001001

DECIPHER ITERATION:  4
DECIPHER:  L = 00101110001111100010010011001110   R = 10111111110
       100001111000001001001
KEY SCH:  K = 101001101111001101010000011001101111011101111111
F:  E[R] = 000101011100000111111100000100001001011001011100
F:  E[R]⊕K = 101100110011001010101100011101100110000100100011
F:  OUTP OF SEL FCN = 00100110001101111000010100100001
F:  OUTP OF PERM P = 11000101010000000011111000011100
DECIPHER:  F(L,K) = 11000101010000000011111000011100
DECIPHER:  R⊕F(L,K) = 01111010100100001100111001010101
DECIPHER:  PREV L = 01111010100100001100111001010101   PREV R = 0
       0101110001111100010010011001110
```

```
DECIPHER ITERATION:  3
DECIPHER:  L = 01111010100100001100111001010101  R = 00101110001
     111100010010011001110
KEY SCH:  K = 111001000101101001110010111111101011100111001011
F:  E[R] = 101111110101010010100001011001011100001010101010
F:  E[R]⊕K = 010110110000111011010011100110110111101101100001
F:  OUTP OF SEL FCN = 11000101010101111011011110100010
F:  OUTP OF PERM P = 10100101111000111111010010011001
DECIPHER:  F(L,K) = 10100101111000111111010010011001
DECIPHER:  R⊕F(L,K) = 10001011110111011101000001010111
DECIPHER:  PREV L = 10001011110111011101000001010111  PREV R = 0
     1111010100100001100111001010101


DECIPHER ITERATION:  2
DECIPHER:  L = 10001011110111011101000001010111  R = 01111010100
     100001100111001010101
KEY SCH:  K = 101000000100101100111001001110111001001111101111
F:  E[R] = 110001010111111101111110111010100000001010101111
F:  E[R]⊕K = 011001011110100010001001100111010010010101000000
F:  OUTP OF SEL FCN = 10011010011001100111110101011101
F:  OUTP OF PERM P = 01111010110111010011100010101110
DECIPHER:  F(L,K) = 01111010110111010011100010101110
DECIPHER:  R⊕F(L,K) = 00000000010011011111011011111011
DECIPHER:  PREV L = 00000000010011011111011011111011  PREV R = 1
     0001011110111011101000001010111


DECIPHER ITERATION:  1
DECIPHER:  L = 00000000010011011111011011111011  R = 10001011110
     111011101000001010111
KEY SCH:  K = 1110000001001011011100110101111111010100111001111
F:  E[R] = 100000000000001001011011111110101101011111110110
F:  E[R]⊕K = 0110000001001010010111101010001010111111000111001
F:  OUTP OF SEL FCN = 01011111110100100101111000000011
F:  OUTP OF PERM P = 01110100011011111100100100011010
DECIPHER:  F(L,K) = 01110100011011111100100100011010
DECIPHER:  R⊕F(L,K) = 11111111101100100001100101001101
DECIPHER:  PREV L = 11111111101100100001100101001101  PREV R = 0
     00000000100110111110110111111011
```

The result of iteration number 1 of the reverse cipher function is a set of two 32-bit blocks, referred to as previous L and previous R. The two blocks are L_0 and R_0, covered earlier for encryption.

The Permuted Input Block

The result of the last iteration–i.e., L_0 and R_0–are catenated to form the original permuted input block. The inverse initial permutation is then applied to the permuted input block yielding the plain text input block:

```
PERMUTTED INPUT BLOCK = 1111111110110010000110010100110100000000
      010011011111011011111011
INPUT BLOCK = 01100111010110100110100101100111010111100101101001
      10101101011010
```

At this stage, the 64-bit plain text is in binary form and would ordinarily be used in the execution of an application program. In this case, it is converted back to character form so that it can be displayed.

Decoding the Plain Text

The last step in the bitwise walk-through of the decryption process is to decode the 64 bits of the plain text into character form. The 64-bit plain text block:

```
0110011101011010011010010110011101011110010110100110101101011010
```

is first divided into 8-bit bytes as follows:

```
01100111 01011010 01101001 01100111 01011110 01011010 01101011 0
      1011010
```

and is then decoded into plain text characters:

Bit Pattern	*Character*
01100111	R
01011010	E
01101001	T
01100111	R
01011110	I
01011010	E
01101011	V
01011010	E

The resulting plain text in character form is:

```
RETRIEVE
```

This concludes the bitwise walk-through. Readers interested in a more detailed trace of the calculations may further modify the functions given in Appendix C.

SELECTED READINGS

Data Encryption Standard, U.S. Department of Commerce, National Bureau of Standards, FIPS publication 46, 1977 January 15.

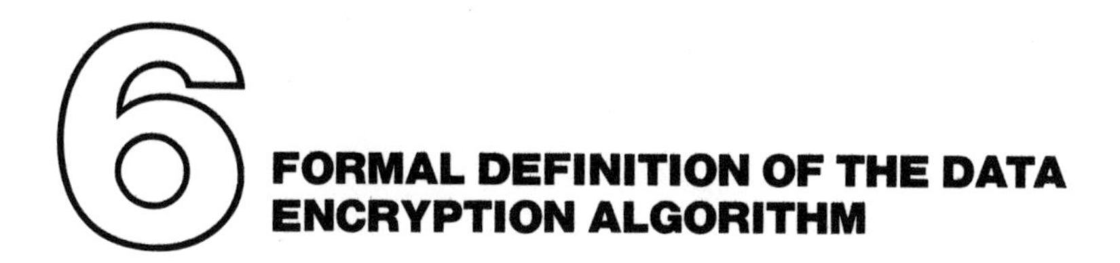

6 FORMAL DEFINITION OF THE DATA ENCRYPTION ALGORITHM

6.1 INTRODUCTION

In 1964, Falkoff, Iverson, and Sussenguth pioneered a unique concept by using the APL language for describing the then new System/360 computer.[1] The objective was to formalize the definition of a complex system using a sufficiently powerful programming language. In fact, the primitives of the APL language have been formally defined by APL functions,[2] using the following criteria for formalization:

1. Primitives are completely and exactly defined for all cases.
2. The functions are executable in APL and are working models.

A similar approach, using the same criteria, is employed here as a means of formalizing the definition of the standard data encryption algorithm. Two main APL functions are involved: CIPHER for encryption and DECIPHER for decryption. Figure 6.1 gives hier-

[1] A. D. Falkoff, K. E. Iverson, and E. Sussenguth, "A Formal Description of System/360," *IBM Systems Journal*, volume 3, number 3 (1964).
[2] R. H. Lathwell, and J. E. Mezei, *A Formal Description of APL* (IBM Philadelphia Scientific Center, Technical Report Number 320-3008, November, 1971).

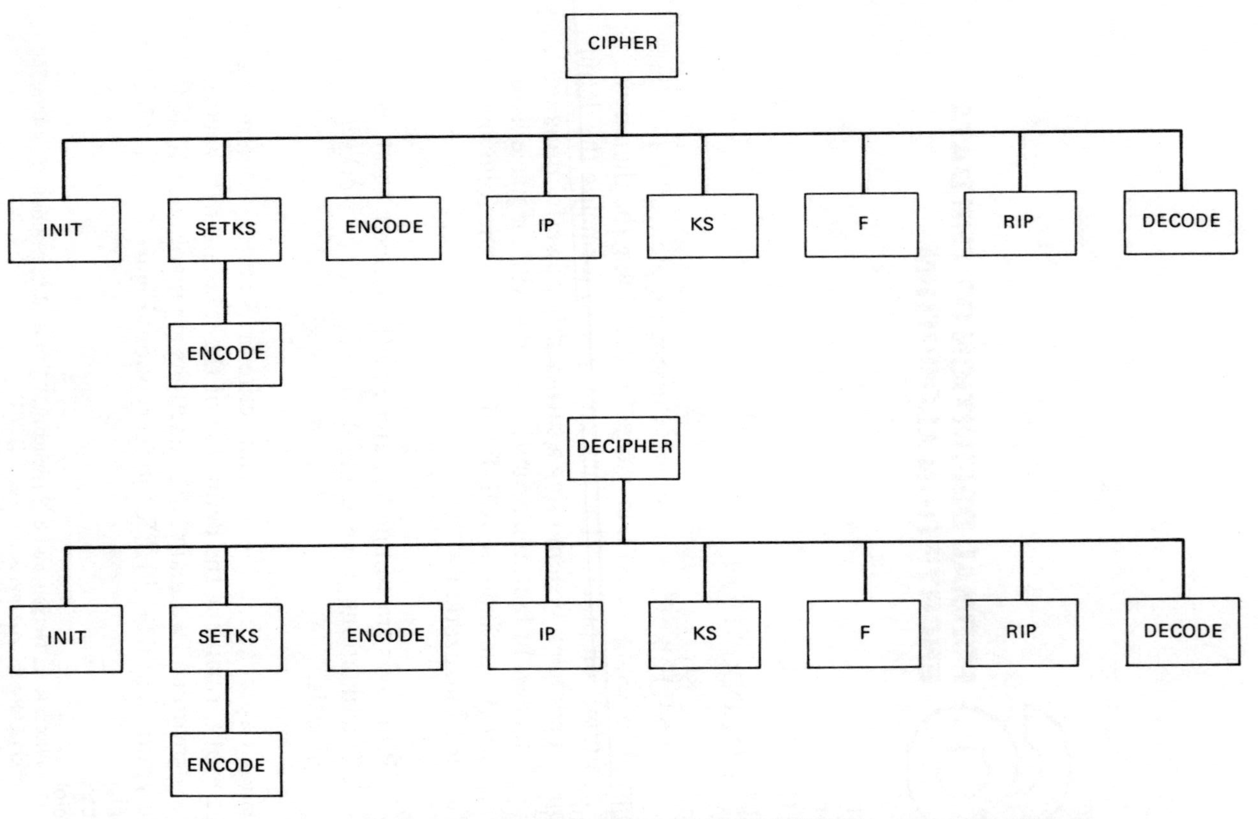

Figure 6.1 Hierarchy diagrams of the CIPHER and DECIPHER functions, denoting the functions that are referenced by these and other functions

archy diagrams of the CIPHER and DECIPHER functions and delineates the other APL functions used in the encryption and decryption processes. The diagrams are intended to present an overview of the APL functions that are referenced during the execution of CIPHER and DECIPHER. The computational operations performed by each of the functions shown in Figure 6.1 are summarized in Table 6.1.

6.2 ILLUSTRATIVE EXAMPLES AND PROCESSING CONSIDERATIONS

Figure 6.2 contains an illustrative example of encryption and decryption using the CIPHER and DECIPHER functions. In the example, the plain text message "ABC 123+" is encrypted using the CIPHER function with the key "COMPUTER". (Note here that the quotation marks are not part of the message or key and serve only as delimiters.) The 64 bits of cipher text are converted back to character form for printing purposes and stored in variable TXT, which is displayed. Not all bit patterns are defined so the resulting cipher text in character form may include undefined symbols.* This is the case in Figure 6.2. The cipher text is then decrypted into the original plain text through a reversal of the process of encryption. This is accomplished with the DECIPHER function.

In a computer, character data are normally stored as bit patterns that are moved internally in parallel. However, when data is transmitted over communications lines or stored on an external

```
      TXT←'ABC 123+' CIPHER 'COMPUTER'
      TXT
∘RO⍹]A⊃⍹
      TXT DECIPHER 'COMPUTER'
ABC 123+
```

Figure 6.2 Illustrative example of encryption and decryption. In this case, the plain text message "ABD 123+" is enciphered using the cipher key "COMPUTER".

*The binary equivalents of APL characters are given in Appendix B.

Table 6.1 Computational operations performed by each of the APL functions used in the formal definition of the standard data encryption algorithm

Function	*Computational Operation Performed*
CIPHER	Performs the data encryption process. Input is the plain text and the cipher key. Output is the cipher text.
DECIPHER	Performs the data decryption process. Input is the cipher text and the cipher key. Output is the original plain text.
ENCODE	Converts APL characters to corresponding logical bit patterns.
DECODE	Converts logical bit patterns to corresponding APL characters.
F	Executes the cipher function $f(A,K)$, where A is either L_i or R_i, and K is the appropriate subkey.
INIT	Sets up the selection and permutation matrices as global arrays.
IP	Performs the initial permutation. Input is the input block. Output is the permuted input block.
KS	Selects the Ith subkey.
RIP	Performs the inverse initial permutation. Input is the preoutput block. Output is the output cipher text in binary form.
SFTKS	Computes a table of 16 subkeys generated from the 64-bit cipher key.

medium, the process usually takes place serially on a bit-by-bit basis. The process of converting from character (or parallel) form to bit-by-bit form is known as *serialization*, which occurs just prior

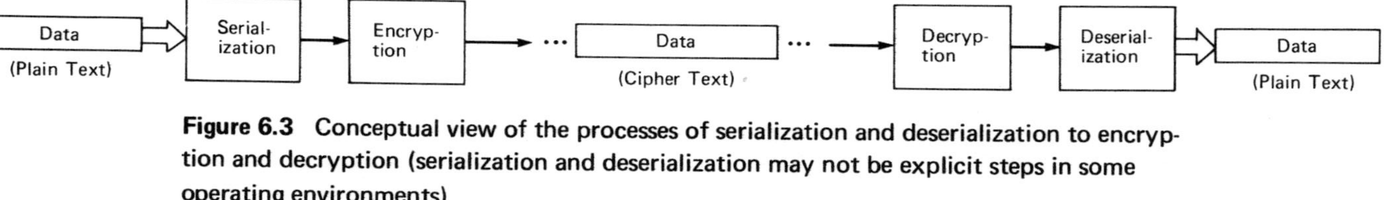

Figure 6.3 Conceptual view of the processes of serialization and deserialization to encryption and decryption (serialization and deserialization may not be explicit steps in some operating environments)

to transmission or storage. At this point in the computing cycle, the character forms of the various bit patterns are not significant. The same philosophy holds true for data reception and retrieval. The actual bit patterns are not significant as long as they can be converted to the original plain text through the processes of decryption and deserialization. Serialization and deserialization are depicted in relation to encryption and decryption in Figure 6.3.

In machine-level programming (i.e., assembler language programming), the facility is normally available for handling the bits that comprise a character, and serialization and deserialization are not customarily regarded as separate steps. In higher-level language programming, however, it is frequently necessary to convert the bits that would comprise a character into logical values before the standard data encryption algorithm can be applied.

Figures 6.4 and 6.5 contain additional illustrative examples. In Figure 6.4, the plain text message "ABC 123+" is encrypted and decrypted using the cipher key "FEBRUARY"; in Figure 6.5, the plain text message "RETRIEVE" is also encrypted and decrypted using the cipher key "FEBRUARY".

```
      TXT←'ABC 123+' CIPHER 'FEBRUARY'
      TXT
&⌹⌹⍦⍹⍥⌹⍋
      TXT DECIPHER 'FEBRUARY'
ABC 123+
```

Figure 6.4 Illustrative example of encryption and decryption. In this case, the plain text message "ABC 123+" is enciphered using the cipher key "FEBRUARY".

```
      TXT←'RETRIEVE' CIPHER 'FEBRUARY'
      TXT
¯⌿⌹⌹V̲⌹ρ⌈
      TXT DECIPHER 'FEBRUARY'
RETRIEVE
```

Figure 6.5 Illustrative example of encryption and decryption. In this case, the plain text message "RETRIEVE" is enciphered using the cipher key "FEBRUARY".

6.3 PROCESSING OVERVIEW

The computations performed in the CIPHER and DECIPHER functions parallel the description of the standard data encryption algorithm given in chapter 4. Figure 6.6 gives flow diagrams of CIPHER and DECIPHER, and Figure 6.7 gives a complete APL listing of the functions that comprise the formal definition of the standard data encryption algorithm.

In APL, the dimensions of arrays and the storage for arrays is maintained dynamically. The input plain text, the output cipher text, and the cipher key are stored as character vectors—each with a dimension of eight characters. After encoding (i.e., serialization), corresponding bit patterns are stored as logical vectors—each with a dimension depending upon the variable stored and the stage of computation.

Full comprehension of the APL functions requires a knowledge of the APL language. However, the functions can be generally understood with the knowledge that APL statements are executed from right to left and a function reference takes one of the following forms:

Type of Function Reference	*Form* (FCN is function name)	*Note*
Niladic	FCN	Takes no arguments; does not return an explicit result.
Monadic	FCN Y	Takes one argument, does not return an explicit result.
Monadic	R←FCN Y	Takes one argument and returns an explicit result.
Dyadic	R←X FCN Y	Takes two arguments and returns an explicit result.

The left arrow (←) denotes replacement, the comma (,) denotes catenation, the up arrow symbol (↑) denotes take, and the down arrow symbol (↓) denotes drop. For example, the statement

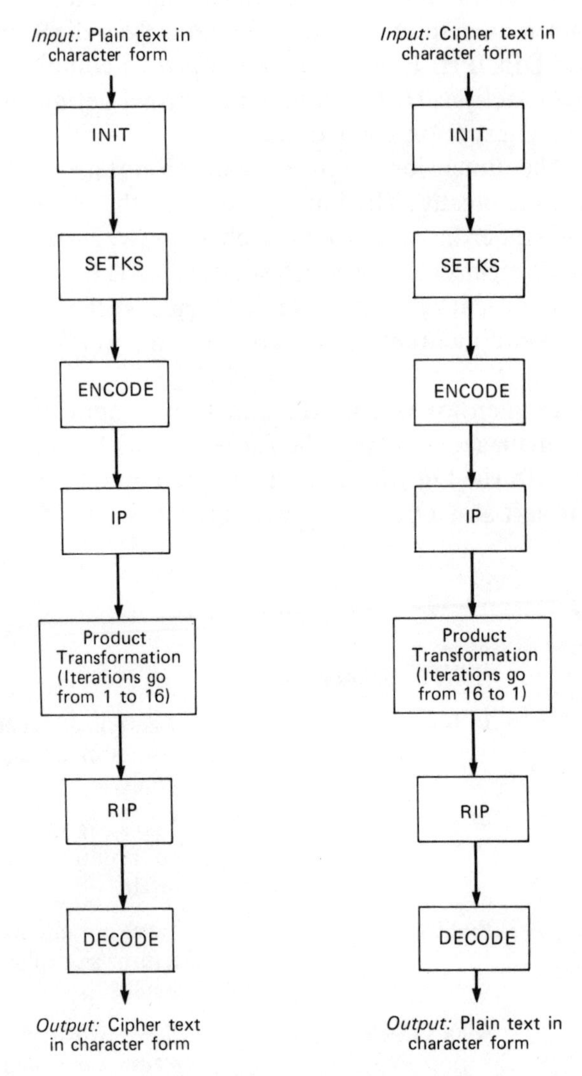

Figure 6.6 Flow diagrams of the CIPHER and DECIPHER functions

```
      ∇CIPHER[⎕]∇
    ∇ R←T CIPHER K;L;R;LN;RN;I
[1]     INIT
[2]     SETKS K
[3]     L←32↑T,R←32↓T←IP ENCODE T
[4]     I←1
[5]   AGN:LN←R
[6]     RN←2|L+R F KS I
[7]     →(16<I←I+1)/OUT
[8]     →AGN,R←32↑RN,L←LN
[9]   OUT:R←DECODE RIP RN,LN
    ∇


      ∇DECIPHER[⎕]∇
    ∇ R←T DECIPHER K;L;R;LN;RN;I
[1]     INIT
[2]     SETKS K
[3]     RN←32↑T,LN←32↓T←IP ENCODE T
[4]     I←16
[5]   AGN:R←LN
[6]     L←2|RN+LN F KS I
[7]     →(0≥I←I-1)/OUT
[8]     →AGN,RN←32↑R,LN←L
[9]   OUT:R←DECODE RIP L,R
    ∇

      ∇DECODE[⎕]∇
    ∇ R←DECODE B;⎕IO
[1]     ⎕IO←0
[2]     R←⎕AV[2⊥⍉ 8 8 ⍴B]
    ∇


      ∇ENCODE[⎕]∇
    ∇ R←ENCODE M;⎕IO
[1]     ⎕IO←0
[2]     R←,⍉(8⍴2)⊤⎕AV⍳M
    ∇


      ∇F[⎕]∇
    ∇ X←R F K;T;I
[1]   ⍝ USES 1 ORIGIN INDEXING
[2]     T← 8 6 ⍴2|R[E]+K
[3]     X←⍳0
[4]     I←1
[5]   LP:X←X, 2 2 2 2 ⊤S[I;1+2⊥(T[I;])[1 6];1+2⊥(T[I;])[2 3 4 5]]
[6]     →(8≥I←I+1)/LP
[7]     X←X[P]
    ∇
```

Figure 6.7 Complete APL listing of the functions in the formal definition of the standard data encryption algorithm

```
      ∇INIT[⎕]∇
    ∇ INIT;S1;S2;S3;S4;S5;S6;S7;S8
[1]   Q← 58 50 42 34 26 18 10 2
[2]   Q←Q, 60 52 44 36 28 20 12 4
[3]   Q←Q, 62 54 46 38 30 22 14 6
[4]   Q←Q, 64 56 48 40 32 24 16 8
[5]   Q←Q, 57 49 41 33 25 17 9 1
[6]   Q←Q, 59 51 43 35 27 19 11 3
[7]   Q←Q, 61 53 45 37 29 21 13 5
[8]   Q←Q, 63 55 47 39 31 23 15 7
[9]   E← 32 1 2 3 4 5
[10]  E←E, 4 5 6 7 8 9
[11]  E←E, 8 9 10 11 12 13
[12]  E←E, 12 13 14 15 16 17
[13]  E←E, 16 17 18 19 20 21
[14]  E←E, 20 21 22 23 24 25
[15]  E←E, 24 25 26 27 28 29
[16]  E←E, 28 29 30 31 32 1
[17]  SHFT← 1 1 2 2 2 2 2 2 1 2 2 2 2 2 2 1
[18]  P← 16 7 20 21
[19]  P←P, 29 12 28 17
[20]  P←P, 1 15 23 26
[21]  P←P, 5 18 31 10
[22]  P←P, 2 8 24 14
[23]  P←P, 32 27 3 9
[24]  P←P, 19 13 30 6
[25]  P←P, 22 11 4 25
[26]  S1← 14 4 13 1 2 15 11 8 3 10 6 12 5 9 0 7
[27]  S1←S1, 0 15 7 4 14 2 13 1 10 6 12 11 9 5 3 8
[28]  S1←S1, 4 1 14 8 13 6 2 11 15 12 9 7 3 10 5 0
[29]  S1←S1, 15 12 8 2 4 9 1 7 4 11 1 14 10 0 6 13
[30]  S2← 15 1 8 14 6 11 3 4 9 7 1 13 12 0 5 10
[31]  S2←S2, 3 13 4 7 15 2 8 14 12 0 1 10 6 9 11 5
[32]  S2←S2, 0 14 7 11 10 4 13 1 5 8 12 6 9 3 2 15
[33]  S2←S2, 13 8 10 1 3 15 4 2 11 6 7 12 0 5 14 9
[34]  S3← 10 0 9 14 6 3 15 5 1 13 12 7 11 4 2 8
[35]  S3←S3, 13 7 0 9 3 4 6 10 2 8 5 14 12 11 15 1
[36]  S3←S3, 13 6 4 9 8 15 3 0 11 1 2 12 5 10 14 7
[37]  S3←S3, 1 10 13 0 6 9 8 7 4 15 14 3 11 5 2 12
[38]  S4← 7 13 14 3 0 6 9 10 1 2 8 5 11 12 4 15
[39]  S4←S4, 13 8 11 5 6 15 0 3 4 7 2 12 1 10 14 9
[40]  S4←S4, 10 6 9 0 12 11 7 13 15 1 3 14 5 2 8 4
[41]  S4←S4, 3 15 0 6 10 1 13 8 9 4 5 11 12 7 2 14
[42]  S5← 2 12 4 1 7 10 11 6 8 5 3 15 13 0 14 9
[43]  S5←S5, 14 11 2 12 4 7 13 1 5 0 15 10 3 9 8 6
[44]  S5←S5, 4 2 1 11 10 13 7 8 15 9 12 5 6 3 0 14
[45]  S5←S5, 11 8 12 7 1 14 2 13 6 15 0 9 10 4 5 3
[46]  S6← 12 1 10 15 9 2 6 8 0 13 3 4 14 7 5 11
[47]  S6←S6, 10 15 4 2 7 12 9 5 6 1 13 14 0 11 3 8
[48]  S6←S6, 9 14 15 5 2 8 12 3 7 0 4 10 1 13 11 6
[49]  S6←S6, 4 3 2 12 9 5 15 10 11 14 1 7 6 0 8 13
[50]  S7← 4 11 2 14 15 0 8 13 3 12 9 7 5 10 6 1
```

Figure 6.7 Continued

```
[51]  S7←S7, 13 0 11 7 4 9 1 10 14 3 5 12 2 15 8 6
[52]  S7←S7, 1 4 11 13 12 3 7 14 10 15 6 8 0 5 9 2
[53]  S7←S7, 6 11 13 8 1 4 10 7 9 5 0 15 14 2 3 12
[54]  S8← 13 2 8 4 6 15 11 1 10 9 3 14 5 0 12 7
[55]  S8←S8, 1 15 13 8 10 3 7 4 12 5 6 11 0 14 9 2
[56]  S8←S8, 7 11 4 1 9 12 14 2 0 6 10 13 15 3 5 8
[57]  S8←S8, 2 1 14 7 4 10 8 13 15 12 9 0 3 5 6 11
[58]  S← 8 4 16 ⍴S1,S2,S3,S4,S5,S6,S7,S8
[59]  PC1A← 57 49 41 33 25 17 9
[60]  PC1A←PC1A, 1 58 50 42 34 26 18
[61]  PC1A←PC1A, 10 2 59 51 43 35 27
[62]  PC1A←PC1A, 19 11 3 60 52 44 36
[63]  PC1B← 63 55 47 39 31 23 15
[64]  PC1B←PC1B, 7 62 54 46 38 30 22
[65]  PC1B←PC1B, 14 6 61 53 45 37 29
[66]  PC1B←PC1B, 21 13 5 28 20 12 4
[67]  PC2← 14 17 11 24 1 5
[68]  PC2←PC2, 3 28 15 6 21 10
[69]  PC2←PC2, 23 19 12 4 26 8
[70]  PC2←PC2, 16 7 27 20 13 2
[71]  PC2←PC2, 41 52 31 37 47 55
[72]  PC2←PC2, 30 40 51 45 33 48
[73]  PC2←PC2, 44 49 39 56 34 53
[74]  PC2←PC2, 46 42 50 36 29 32
    ∇


       ∇IP[⎕]∇
    ∇ R←IP L
[1]    R←L[Q]
    ∇


       ∇KS[⎕]∇
    ∇ R←KS N
[1]    R←KN[N;]
    ∇


       ∇RIP[⎕]∇
    ∇ R←RIP L
[1]    R←L[Q⍳⍳64]
    ∇


       ∇SETKS[⎕]∇
    ∇ SETKS K;KB;C;D;I
[1]    KN← 16 48 ⍴0
[2]    KB←ENCODE K
[3]    I←1
[4]    C←KB[PC1A]
[5]    D←KB[PC1B]
[6]  LP:C←(1↑(I-1)⌽SHFT)⌽C
[7]    D←(1↑(I-1)⌽SHFT)⌽D
[8]    KN[I;]←(C,D)[PC2]
[9]    →(16≥I←I+1)/LP
    ∇
```

Figure 6.7 Continued

```
L←32↑C,R←32↓C←A,B
```

indicates that the following operations should be performed:

1. Catenate B with A.
2. Replace C with the result of step 1.
3. Drop the first 32 characters of C (C is not changed) and replace R (for right) with the remaining characters.
4. Catenate R with C.
5. Take the first 32 characters of C,R.
6. Replace L (for left) with the result of step 5.

In the APL functions, some statements are less complex than the one given above and others require an in-depth knowledge of APL. The reader is directed to the references for books on the APL language.

The remaining sections in this chapter cover the various functions in the formal definition.

6.4 ENCRYPTION

Encryption is performed with the function CIPHER, listed in Figure 6.8. Statements numbered 1 and 2 call the INIT and SETKS functions to initialize the permutation and selection matrices and to establish the table of subkeys, respectively. Statement numbered 3, listed as follows,

```
L←32↑T,R↓32←T IP ENCODE T
```

encodes the plain text T into a 64-bit block, performs the initial permutation yielding the permuted input block, and then divides the permuted input block into left and right blocks of 32 bits each. The left and right blocks are L_0 and R_0, respectively.

Statements numbered 5 through 8 perform one iteration of the product transformation, and the iteration is performed as control

```
        ∇CIPHER[⎕]∇
      ∇ R←T CIPHER K;L;R;LN;RN;I
[1]     INIT
[2]     SETKS K
[3]     L←32↑T,R←32↓T←IP ENCODE T
[4]     I←1
[5]   AGN:LN←R
[6]     RN←2|L+R F KS I
[7]     →(16<I←I+1)/OUT
[8]     →AGN,R←32↑RN,L←LN
[9]   OUT:R←DECODE RIP RN,LN
      ∇
```

Figure 6.8 Listing of the APL function CIPHER (this function performs encryption)

variable I goes from 1 to 16. Statement numbered 6 is an important one. It is listed as follows:

RN←2|L+R F KS I

The statement performs the following computations, as listed in chapter 4:

$R_n \leftarrow L_{n-1} \oplus f(R_{n-1},K_n)$

After 16 iterations, control is passed to statement numbered 9, which forms the preoutput block and then successively performs the inverse initial permutation and the decoding of the 64-bit cipher text into character form.

6.5 DECRYPTION

Decryption is performed with the function DECIPHER, listed in Figure 6.9. The DECIPHER function is essentially the same as the CIPHER function, except that the product transformation is performed in reverse order. Statements numbered 1 and 2 call the INIT and SETKS functions to initialize the permutation and selection matrices and to establish the table of subkeys, respectively. Statement numbered 3, listed as follows,

RN←32↑T,LN←32↓T←IP ENCODE T

encodes the cipher text T into a 64-bit block, performs the initial permutation* yielding the preoutput block, and then reverses the block transformation that produced the preoutput block. The result is divided into left and right blocks of 32-bits each, which are L_{16} and R_{16}, respectively.

```
        ∇DECIPHER[⎕]∇
      ∇ R←T DECIPHER K;L;R;LN;RN;I
[1]     INIT
[2]     SETKS K
[3]     RN←32↑T,LN←32↓T←IP ENCODE T
[4]     I←16
[5]    AGN:R←LN
[6]     L←2|RN+LN F KS I
[7]     →(0≥I←I-1)/OUT
[8]     →AGN,RN←32↑R,LN←L
[9]    OUT:R←DECODE RIP L,R
      ∇
```

Figure 6.9 Listing of the APL function DECIPHER (this function performs decryption)

Statements numbered 5 through 8 perform one iteration of the product transformation, and the iteration is performed as control variable I goes from 16 to 1. Statement numbered 6 is significant and is listed as follows:

L←2|RN+LN F KS I

The statement performs the following computations, as listed in chapter 4:

$$L_{n-1} \leftarrow R_n \oplus f(L_n, K_n)$$

*It should be remembered here that the initial permutation and the inverse initial permutation are exact inverses of each other, so that the initial permutation produces the inverse of the inverse initial permutation in the same way that the inverse initial permutation produces the inverse of the initial permutation.

After 16 iterations, control is passed to statement numbered 9, which forms the original permuted input block by catenating L_0 and R_0—the result of the last iteration—and then successively performs the inverse initial permutation—yielding the input block—and the decoding of the 64-bit plain text into character form.

6.6 ENCODING AND DECODING

The encoding and decoding functions, referred to earlier as serialization and deserialization, convert an eight-character block to a 64-bit block and also reverse the process, respectively. The ENCODE and DECODE functions are listed in Figures 6.10 and 6.11, respectively.

```
          ∇ENCODE[⎕]∇
        ∇ R←ENCODE M;⎕IO
[1]       ⎕IO←0
[2]       R←,⍉(8⍴2)⊤⎕AV⍳M
        ∇
```

Figure 6.10 Listing of the APL function ENCODE (this function converts an 8-character block to a 64-bit block)

```
          ∇DECODE[⎕]∇
        ∇ R←DECODE B;⎕IO
[1]       ⎕IO←0
[2]       R←⎕AV[2⊥⍉ 8 8 ⍴B]
        ∇
```

Figure 6.11 Listing of the APL function DECODE (this function converts a 64-bit block to an 8-character block)

6.7 THE CIPHER FUNCTION

The cipher function is represented by the APL function named F and is listed in Figure 6.12. Statement numbered 2 modulo-2 adds the key K to selection E applied to the input argument R. The result of statement numbered 2 is divided into 8 groups of 6 bits.

```
      ∇F[⎕]∇
    ∇ X←R F K;T;I
[1]   ⍝ USES 1 ORIGIN INDEXING
[2]    T← 8 6 ⍴2|R[E]+K
[3]    X←⍳0
[4]    I←1
[5]   LP:X←X, 2 2 2 2 ⊤S[I;1+2⊥(T[I;])[1 6];1+2⊥(T[I;])[2 3 4 5]]
[6]    →(8≥I←I+1)/LP
[7]    X←X[P]
    ∇
```

Figure 6.12 Listing of the APL function F (this function performs the cipher function)

Statements numbered 5 and 6 perform the selection function S_n, as n runs from 1 to 8. The result is catenated into a logical vector. Statement numbered 7 applies to permutation P producing the output of the function F.

6.8 INITIAL AND INVERSE INITIAL PERMUTATIONS

The initial permutation (IP) and the inverse initial permutation (RIP) are listed in Figures 6.13 and 6.14, respectively. In the initial permutation, the statement

```
R←L[Q]
```

selects the bits in the input block L according to the initial permutation matrix Q. The output is the plain text in permuted form (during encryption) or the cipher text in permuted form (during decryption).

```
      ∇IP[⎕]∇
    ∇ R←IP L
[1]    R←L[Q]
    ∇
```

Figure 6.13 Listing of the APL function IP (this function performs the initial permutation)

```
        ∇RIP[⎕]∇
      ∇ R←RIP L
[1]     R←L[Q⍳⍳64]
      ∇
```

Figure 6.14 Listing of the APL function RIP (this function performs the inverse initial permutation)

In the inverse initial permutation, the statement

R←L[Q⍳⍳64]

first forms a vector of integers (⍳64) from 1 to 64 which represents the bits in the original plain text. Then, the index of operation Q⍳⍳64 selects the position (or index) of those elements in the initial permutation matrix. The final selection operation L[Q⍳⍳64] selects the elements of the argument L in an order that reverses the initial permutation. The output is the cipher text (during encryption) or the plain text (during decryption).

6.9 THE KEY SCHEDULE

The key schedule function (KS) is used during the execution of the cipher function (F) to retrieve the Ith subkey during the Ith iteration of the product transformation. The function KS is listed in Figure 6.15. Input to KS is the index I. Output from KS is the 48-bit subkey that was generated during the Ith iteration of the function SETKS, which established the table of subkeys during initialization.

```
        ∇KS[⎕]∇
      ∇ R←KS N
[1]     R←KN[N;]
      ∇
```

Figure 6.15 Listing of the APL function KS (this function selects the Ith subkey from the key schedule)

6.10 INITIALIZATION

Two APL functions are used for initialization: INIT and SETKS. The function INIT, listed in Figure 6.16, establishes the permutation matrices given in chapter 5. In INIT, an attempt is made to present the indices in the same order given in chapter 4, so that they can be recognized by the reader. The permutation matrices generated are:

Matrix	*Permutation*
Q	Initial permutation
E	Used in the cipher function
SHFT	Used to calculate the key schedule
P	Used in the cipher function
S	Selection functions
PC1A	Permuted choice 1-C_0
PC1B	Permuted choice 1-D_0
PC2	Permuted choice 2

```
          ∇INIT[⎕]∇
        ∇ INIT;S1;S2;S3;S4;S5;S6;S7;S8
[1]       Q← 58 50 42 34 26 18 10 2
[2]       Q←Q, 60 52 44 36 28 20 12 4
[3]       Q←Q, 62 54 46 38 30 22 14 6
[4]       Q←Q, 64 56 48 40 32 24 16 8
[5]       Q←Q, 57 49 41 33 25 17 9 1
[6]       Q←Q, 59 51 43 35 27 19 11 3
[7]       Q←Q, 61 53 45 37 29 21 13 5
[8]       Q←Q, 63 55 47 39 31 23 15 7
[9]       E← 32 1 2 3 4 5
[10]      E←E, 4 5 6 7 8 9
[11]      E←E, 8 9 10 11 12 13
[12]      E←E, 12 13 14 15 16 17
[13]      E←E, 16 17 18 19 20 21
[14]      E←E, 20 21 22 23 24 25
[15]      E←E, 24 25 26 27 28 29
[16]      E←E, 28 29 30 31 32 1
[17]      SHFT← 1 1 2 2 2 2 2 2 1 2 2 2 2 2 2 1
[18]      P← 16 7 20 21
[19]      P←P, 29 12 28 17
[20]      P←P, 1 15 23 26
[21]      P←P, 5 18 31 10
[22]      P←P, 2 8 24 14
[23]      P←P, 32 27 3 9
[24]      P←P, 19 13 30 6
[25]      P←P, 22 11 4 25
```

Figure 6.16 Listing of the APL function INIT (this function establishes the permutation matrices)

```
[26]  S1← 14 4 13 1 2 15 11 8 3 10 6 12 5 9 0 7
[27]  S1←S1, 0 15 7 4 14 2 13 1 10 6 12 11 9 5 3 8
[28]  S1←S1, 4 1 14 8 13 6 2 11 15 12 9 7 3 10 5 0
[29]  S1←S1, 15 12 8 2 4 9 1 7 4 11 1 14 10 0 6 13
[30]  S2← 15 1 8 14 6 11 3 4 9 7 1 13 12 0 5 10
[31]  S2←S2, 3 13 4 7 15 2 8 14 12 0 1 10 6 9 11 5
[32]  S2←S2, 0 14 7 11 10 4 13 1 5 8 12 6 9 3 2 15
[33]  S2←S2, 13 8 10 1 3 15 4 2 11 6 7 12 0 5 14 9
[34]  S3← 10 0 9 14 6 3 15 5 1 13 12 7 11 4 2 8
[35]  S3←S3, 13 7 0 9 3 4 6 10 2 8 5 14 12 11 15 1
[36]  S3←S3, 13 6 4 9 8 15 3 0 11 1 2 12 5 10 14 7
[37]  S3←S3, 1 10 13 0 6 9 8 7 4 15 14 3 11 5 2 12
[38]  S4← 7 13 14 3 0 6 9 10 1 2 8 5 11 12 4 15
[39]  S4←S4, 13 8 11 5 6 15 0 3 4 7 2 12 1 10 14 9
[40]  S4←S4, 10 6 9 0 12 11 7 13 15 1 3 14 5 2 8 4
[41]  S4←S4, 3 15 0 6 10 1 13 8 9 4 5 11 12 7 2 14
[42]  S5← 2 12 4 1 7 10 11 6 8 5 3 15 13 0 14 9
[43]  S5←S5, 14 11 2 12 4 7 13 1 5 0 15 10 3 9 8 6
[44]  S5←S5, 4 2 1 11 10 13 7 8 15 9 12 5 6 3 0 14
[45]  S5←S5, 11 8 12 7 1 14 2 13 6 15 0 9 10 4 5 3
[46]  S6← 12 1 10 15 9 2 6 8 0 13 3 4 14 7 5 11
[47]  S6←S6, 10 15 4 2 7 12 9 5 6 1 13 14 0 11 3 8
[48]  S6←S6, 9 14 15 5 2 8 12 3 7 0 4 10 1 13 11 6
[49]  S6←S6, 4 3 2 12 9 5 15 10 11 14 1 7 6 0 8 13
[50]  S7← 4 11 2 14 15 0 8 13 3 12 9 7 5 10 6 1
[51]  S7←S7, 13 0 11 7 4 9 1 10 14 3 5 12 2 15 8 6
[52]  S7←S7, 1 4 11 13 12 3 7 14 10 15 6 8 0 5 9 2
[53]  S7←S7, 6 11 13 8 1 4 10 7 9 5 0 15 14 2 3 12
[54]  S8← 13 2 8 4 6 15 11 1 10 9 3 14 5 0 12 7
[55]  S8←S8, 1 15 13 8 10 3 7 4 12 5 6 11 0 14 9 2
[56]  S8←S8, 7 11 4 1 9 12 14 2 0 6 10 13 15 3 5 8
[57]  S8←S8, 2 1 14 7 4 10 8 13 15 12 9 0 3 5 6 11
[58]  S← 8 4 16 ρS1,S2,S3,S4,S5,S6,S7,S8
[59]  PC1A← 57 49 41 33 25 17 9
[60]  PC1A←PC1A, 1 58 50 42 34 26 18
[61]  PC1A←PC1A, 10 2 59 51 43 35 27
[62]  PC1A←PC1A, 19 11 3 60 52 44 36
[63]  PC1B← 63 55 47 39 31 23 15
[64]  PC1B←PC1B, 7 62 54 46 38 30 22
[65]  PC1B←PC1B, 14 6 61 53 45 37 29
[66]  PC1B←PC1B, 21 13 5 28 20 12 4
[67]  PC2← 14 17 11 24 1 5
[68]  PC2←PC2, 3 28 15 6 21 10
[69]  PC2←PC2, 23 19 12 4 26 8
[70]  PC2←PC2, 16 7 27 20 13 2
[71]  PC2←PC2, 41 52 31 37 47 55
[72]  PC2←PC2, 30 40 51 45 33 48
[73]  PC2←PC2, 44 49 39 56 34 53
[74]  PC2←PC2, 46 42 50 36 29 32
    ∇
```

Figure 6.16 Continued

The function SETKS, listed in Figure 6.17, computes the key schedule comprised of sixteen 48-bit subkeys. The subkeys are stored as KN, which is a 16 × 48 logical matrix established in statement numbered 1. Statement numbered 2 encodes the cipher key K, and statements numbered 4 and 5 compute C_0 and D_0, respectively, by applying permuted choice 1. Statements numbered 6 through 9 go through the 16 iterations necessary to compute the 16 subkeys. Statements numbered 6 and 7 perform the circular left shift operations on C_{i-1} and D_{i-1}, respectively, to yield C_i and D_i. Statement numbered 8 taps off C_i and D_i, catenates them, and then performs permuted choice 2 to yield the ith subkey.

```
          ∇SETKS[⎕]∇
        ∇ SETKS K;KB;C;D;I
[1]       KN← 16 48 ⍴0
[2]       KB←ENCODE K
[3]       I←1
[4]       C←KB[PC1A]
[5]       D←KB[PC1B]
[6]      LP:C←(1↑(I-1)⌽SHFT)⌽C
[7]       D←(1↑(I-1)⌽SHFT)⌽D
[8]       KN[I;]←(C,D)[PC2]
[9]       →(16≥I←I+1)/LP
        ∇
```

Figure 6.17 Listing of the APL function SETKS (this function computes the key schedule)

6.11 PRACTICAL CONSIDERATIONS

The APL functions given in this chapter satisfy the criteria for formalization. Because the APL language is interpreted, however, they do not constitute an efficient method of implementation. In general, the use of compiled code is preferred and the use of either assembler language or a multipurpose programming language, such as PL/I, would be the most efficient means of programming the algorithm.

The examples given in this chapter were run on an IBM 5100 portable computer.

SELECTED READINGS

Data Encryption Standard, U.S. Department of Commerce, National Bureau of Standards, FIPS Publication 46, 1977 January 15.

Katzan, H. *APL User's Guide.* New York: Van Nostrand Reinhold Company, 1971.

——. *The IBM 5100 Portable Computer: A Comprehensive Guide for Users and Programmers.* New York: Van Nostrand Reinhold Company, 1977.

APPENDICES

APPENDIX A

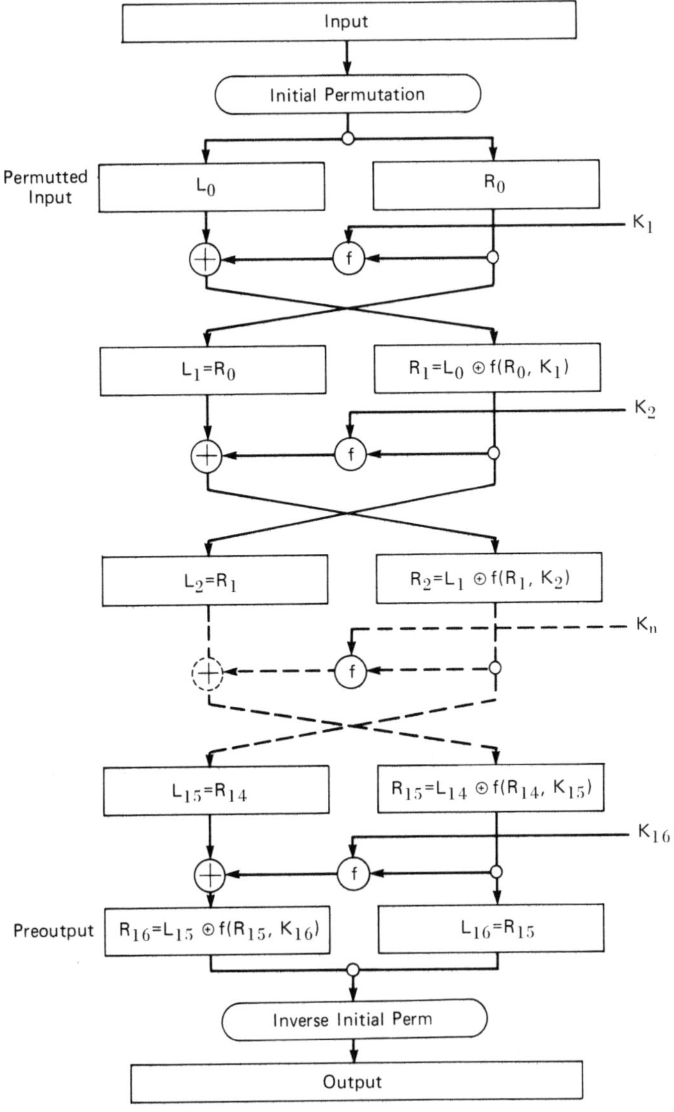

Figure A Diagram of the enciphering computation (reproduced from Data Encryption Standard, FIPS pub 46, p. 8)

APPENDIX B

Index *⎕IO←1*	*Character*	*Bit Pattern*
1	⌷	00000001
2	⌷	00000010
3	⌷	00000011
4	⌷	00000100
5	⌷	00000101
6	⌷	00000110
7	⌷	00000111
8	⌷	00001000
9	⌷	00001001
10	⌷	00001010
11	⌷	00001011
12	⌷	00001100
13	⌷	00001101
14	[	00001110
15	]	00001111
16	(	00010000
17	)	00010001
18	;	00010010
19	/	00010011
20	\	00010100
21	←	00010101
22	→	00010110
23	⌷	00010111
24	⌷	00011000
25	¨	00011001
26	+	00011010
27	−	00011011
28	×	00011100
29	÷	00011101
30	⋆	00011110
31	⌈	00011111
32	⌊	00100000
33	|	00100001
34	∧	00100010
35	∨	00100011
36	<	00100100
37	≤	00100101
38	=	00100110
39	≥	00100111
40	>	00101000
41	≠	00101001
42	α	00101010
43	∊	00101011
44	⍳	00101100
45	⍴	00101101
46	⍵	00101110
47	,	00101111

Figure B APL characters with corresponding bit patterns based on each character's index in the APL atomic vector

Index ⎕*IO←I*	*Character*	*Bit Pattern*
48	!	00110000
49	⌽	00110001
50	⊥	00110010
51	⊤	00110011
52	○	00110100
53	?	00110101
54	∼	00110110
55	↑	00110111
56	↓	00111000
57	⊂	00111001
58	⊃	00111010
59	∩	00111011
60	∪	00111100
61	_	00111101
62	⍝	00111110
63	⌶	00111111
64	∘	01000000
65	⎕	01000001
66	⍞	01000010
67	⍟	01000011
68	⍲	01000100
69	⍱	01000101
70	⍋	01000110
71	⍒	01000111
72	⍉	01001000
73	⊖	01001001
74	⌿	01001010
75	⍀	01001011
76	⌹	01001100
77	⍕	01001101
78	⍎	01001110
79	&	01001111
80	@	01010000
81	#	01010001
82	$	01010010
83	⍙	01010011
84	⍙	01010100
85	⍙	01010101
86	A	01010110
87	B	01010111
88	C	01011000
89	D	01011001
90	E	01011010
91	F	01011011
92	G	01011100
93	H	01011101
94	I	01011110
95	J	01011111
96	K	01100000
97	L	01100001
98	M	01100010
99	N	01100011
100	O	01100100
101	P	01100101
102	Q	01100110
103	R	01100111
104	S	01101000
105	T	01101001
106	U	01101010

Figure B Continued

Index ⎕*IO*←*1*	*Character*	*Bit Pattern*
107	V	01101011
108	W	01101100
109	X	01101101
110	Y	01101110
111	Z	01101111
112	∆	01110000
113	A̲	01110001
114	B̲	01110010
115	C̲	01110011
116	D̲	01110100
117	E̲	01110101
118	F̲	01110110
119	G̲	01110111
120	H̲	01111000
121	I̲	01111001
122	J̲	01111010
123	K̲	01111011
124	L̲	01111100
125	M̲	01111101
126	N̲	01111110
127	O̲	01111111
128	P̲	10000000
129	Q̲	10000001
130	R̲	10000010
131	S̲	10000011
132	T̲	10000100
133	U̲	10000101
134	V̲	10000110
135	W̲	10000111
136	X̲	10001000
137	Y̲	10001001
138	Z̲	10001010
139	∆̲	10001011
140	0	10001100
141	1	10001101
142	2	10001110
143	3	10001111
144	4	10010000
145	5	10010001
146	6	10010010
147	7	10010011
148	8	10010100
149	9	10010101
150	.	10010110
151	¯	10010111
152		10011000
153	'	10011001
154	:	10011010
155	∇	10011011
156		10011100
157		10011101
158		10011110
159		10011111
160	⍫	10100000
161	⌷	10100001
162	⌷	10100010
163	⌷	10100011
164	⌷	10100100
165	⌷	10100101

Figure B Continued

Index ⎕*IO←1*	*Character*	*Bit Pattern*
166	⍞	10100110
167	⍞	10100111
168	⍞	10101000
169	⍞	10101001
170	¬	10101010
171	¨	10101011
172	%	10101100
173	⍙	10101101
174	⍟	10101110
175	Ä	10101111
176	Ö	10110000
177	Ü	10110001
178	Å	10110010
179	Æ	10110011
180	₧	10110100
181	Ñ	10110101
182	£	10110110
183	Ç	10110111
184	Ö	10111000
185	Ä	10111001

⍞ DENOTES AN UNDEFINED CHARACTER
ELEMENTS (186-255) ARE UNUSED BUT EXIST AS BINARY
  BIT PATTERNS

Figure B Continued

APPENDIX C

```
        ∇CIPHER[⎕]∇
      ∇ R←T CIPHER K;L;R;LN;RN;I
[1]     INIT
[2]     SETKS K
[3]     T1←ENCODE T
[4]     T←IP T1
[5]     ('INPUT BLOCK = '),(1 0 ⍕T1)
[6]     ('PERMUTTED INPUT BLOCK = '),(1 0 ⍕T)
[7]     L←32↑T,R←32↓T
[8]     SPACE 10
[9]     ('L(ZERO) = '),(1 0 ⍕L),('  R(ZERO) = '),(1 0 ⍕R)
[10]    SPACE 10
[11]    I←1
[12]  AGN:('CIPHER ITERATION: '),(2 0 ⍕I)
[13]    ('CIPHER:  L = '),(1 0 ⍕L),('  R = '),(1 0 ⍕R)
[14]    LN←R
[15]    T1←R F KS I
[16]    ('CIPHER:  F(R,K) = '),(1 0 ⍕T1)
[17]    T2←2|L+T1
[18]    ('CIPHER:  L⊕F(R,K) = '),(1 0 ⍕T2)
[19]    RN←T2
[20]    ('CIPHER: NEXT L = '),(1 0 ⍕LN),('  NEXT R = '),(1 0 ⍕RN)
[21]    SPACE 1
[22]    →(16<I←I+1)/OUT
[23]    →AGN,R←32↑RN,L←LN
[24]  OUT:T3←RN,LN
[25]    SPACE 10
[26]    ('PREOUTPUT BLOCK = '),(1 0 ⍕T3)
[27]    T4←RIP T3
[28]    ('PERMUTTED OUTPUT BLOCK = '),(1 0 ⍕T4)
[29]    SPACE 10
[30]    R←DECODE T4
      ∇
```

```
        ∇DECIPHER[⎕]∇
      ∇ R←T DECIPHER K;L;R;LN;RN;I
[1]     INIT
[2]     SETKS K
[3]     T1←ENCODE T
[4]     T←IP T1
[5]     ('OUTPUT BLOCK = '),(1 0 ⍕T1)
[6]     ('PREOUTPUT BLOCK = '),(1 0 ⍕T)
[7]     RN←32↑T,LN←32↓T
[8]     SPACE 10
[9]     ('L(16) = '),(1 0 ⍕LN),('  R(16) = '),(1 0 ⍕RN)
[10]    SPACE 10
[11]    I←16
[12]  AGN:('DECIPHER ITERATION: '),(2 0 ⍕I)
[13]    ('DECIPHER:  L = '),(1 0 ⍕LN),('  R = '),(1 0 ⍕RN)
```

Figure C Listing of modified functions used to generate the bit-wise walk-through in chapter 5

```
[14]   R←LN
[15]   T1←LN F KS I
[16]   ('DECIPHER:  F(L,K) = '),(1 0 ⍕T1)
[17]   T2←2|RN+T1
[18]   ('DECIPHER:  R⊕F(L,K) = '),(1 0 ⍕T2)
[19]   L←T2
[20]   ('DECIPHER:  PREV L = '),(1 0 ⍕L),('  PREV R = '),(1 0 ⍕R)
[21]   SPACE 1
[22]   →(0≥I←I-1)/OUT
[23]   →AGN,RN←32↑R,LN←L
[24]  OUT:T3←L,R
[25]   SPACE 10
[26]   ('PERMUTTED INPUT BLOCK = '),(1 0 ⍕T3)
[27]   T4←RIP T3
[28]   ('INPUT BLOCK = '),(1 0 ⍕T4)
[29]   SPACE 10
[30]   B←DECODE T4
     ∇


       ∇DECODE[⎕]∇
     ∇ R←DECODE B;⎕IO
[1]    ⎕IO←0
[2]    R←⎕AV[2⊥⍉ 8 8 ⍴B]
[3]    SPACE 10
[4]    1 0 ⍕B
[5]    SPACE 10
[6]    U←71⍴ 1 1 1 1 1 1 1 1 0
[7]    U\ 1 0 ⍕B
[8]    SPACE 10
[9]    B← 8 8 ⍴B
[10]   I←0
[11]  AGN:(1 0 ⍕B[I;]),(5⍴' '),⎕AV[2⊥,B[I;]]
[12]   →(7≥I←I+1)/AGN
[13]   SPACE 10
     ∇


       ∇ENCODE[⎕]∇
     ∇ R←ENCODE M;⎕IO;T;U;V
[1]    ⎕IO←0
[2]    R←,⍉(8⍴2)⊤⎕AV⍳M
[3]    SPACE 10
[4]    (8 1 ⍴M),(8 5 ⍴' '),(1 0 ⍕T←⍉(8⍴2)⊤⎕AV⍳M)
[5]    SPACE 10
[6]    U←71⍴ 1 1 1 1 1 1 1 1 0
[7]    U\ 1 0 ⍕,T
[8]    SPACE 10
[9]    1 0 ⍕,T
[10]   SPACE 10
     ∇


       ∇F[⎕]∇
     ∇ X←R F K;T;I
[1]    ⍝ USES 1 ORIGIN INDEXING
[2]    T1←R[E]
[3]    ('F:  E[R] = '),(1 0 ⍕T1)
[4]    T2←2|T1+K
[5]    ('F:  E[R]⊕K = '),(1 0 ⍕T2)
[6]    T← 8 6 ⍴T2
[7]    X←⍳0
```

Figure C Continued

```
[8]    I←1
[9]   LP:X←X, 2 2 2 2 ⊤S[I;1+2⊥(T[I;])[1 6];1+2⊥(T[I;])[2 3 4 5]]
[10]   →(8≥I←I+1)/LP
[11]   ('F:  OUTP OF SEL FCN = '),(1 0 ⍕X)
[12]   X←X[P]
[13]   ('F:  OUTP OF PERM P = '),(1 0 ⍕X)
     ∇


       ∇IP[⎕]∇
     ∇ R←IP L
[1]    R←L[Q]
     ∇


       ∇KS[⎕]∇
     ∇ R←KS N
[1]    R←KN[N;]
[2]    ('KEY SCH:  K = '),(1 0 ⍕R)
     ∇


       ∇RIP[⎕]∇
     ∇ R←RIP L
[1]    R←L[Q⍳⍳64]
     ∇


       ∇SETKS[⎕]∇
     ∇ SETKS K;KB;C;D;I
[1]    SPACE 10
[2]    KN← 16 48 ⍴0
[3]    KB←ENCODE K
[4]    I←1
[5]    C←KB[PC1A]
[6]    D←KB[PC1B]
[7]   LP:C←(1↑(I-1)⌽SHFT)⌽C
[8]    D←(1↑(I-1)⌽SHFT)⌽D
[9]    KN[I;]←(C,D)[PC2]
[10]   (' N = '),(2 0 ⍕I),('  C = '),(1 0 ⍕C),('  D = '),(1 0 ⍕D)
[11]   (' SUBKEY = '),(1 0 ⍕KN[I;])
[12]   ' '
[13]   →(16≥I←I+1)/LP
[14]   SPACE 10
     ∇


       ∇SPACE[⎕]∇
     ∇ SPACE N;⎕IO
[1]    ⎕IO←1
[2]    N⍴⎕AV[157]
     ∇


       ∇INIT[⎕]∇
     ∇ INIT;S1;S2;S3;S4;S5;S6;S7;S8
[1]    Q← 58 50 42 34 26 18 10 2
[2]    Q←Q, 60 52 44 36 28 20 12 4
[3]    Q←Q, 62 54 46 38 30 22 14 6
[4]    Q←Q, 64 56 48 40 32 24 16 8
[5]    Q←Q, 57 49 41 33 25 17 9 1
```

Figure C Continued

```
[6]    Q←Q, 59 51 43 35 27 19 11 3
[7]    Q←Q, 61 53 45 37 29 21 13 5
[8]    Q←Q, 63 55 47 39 31 23 15 7
[9]    E← 32 1 2 3 4 5
[10]   E←E, 4 5 6 7 8 9
[11]   E←E, 8 9 10 11 12 13
[12]   E←E, 12 13 14 15 16 17
[13]   E←E, 16 17 18 19 20 21
[14]   E←E, 20 21 22 23 24 25
[15]   E←E, 24 25 26 27 28 29
[16]   E←E, 28 29 30 31 32 1
[17]   SHFT← 1 1 2 2 2 2 2 2 1 2 2 2 2 2 2 1
[18]   P← 16 7 20 21
[19]   P←P, 29 12 28 17
[20]   P←P, 1 15 23 26
[21]   P←P, 5 18 31 10
[22]   P←P, 2 8 24 14
[23]   P←P, 32 27 3 9
[24]   P←P, 19 13 30 6
[25]   P←P, 22 11 4 25
[26]   S1← 14 4 13 1 2 15 11 8 3 10 6 12 5 9 0 7
[27]   S1←S1, 0 15 7 4 14 2 13 1 10 6 12 11 9 5 3 8
[28]   S1←S1, 4 1 14 8 13 6 2 11 15 12 9 7 3 10 5 0
[29]   S1←S1, 15 12 8 2 4 9 1 7 4 11 1 14 10 0 6 13
[30]   S2← 15 1 8 14 6 11 3 4 9 7 1 13 12 0 5 10
[31]   S2←S2, 3 13 4 7 15 2 8 14 12 0 1 10 6 9 11 5
[32]   S2←S2, 0 14 7 11 10 4 13 1 5 8 12 6 9 3 2 15
[33]   S2←S2, 13 8 10 1 3 15 4 2 11 6 7 12 0 5 14 9
[34]   S3← 10 0 9 14 6 3 15 5 1 13 12 7 11 4 2 8
[35]   S3←S3, 13 7 0 9 3 4 6 10 2 8 5 14 12 11 15 1
[36]   S3←S3, 13 6 4 9 8 15 3 0 11 1 2 12 5 10 14 7
[37]   S3←S3, 1 10 13 0 6 9 8 7 4 15 14 3 11 5 2 12
[38]   S4← 7 13 14 3 0 6 9 10 1 2 8 5 11 12 4 15
[39]   S4←S4, 13 8 11 5 6 15 0 3 4 7 2 12 1 10 14 9
[40]   S4←S4, 10 6 9 0 12 11 7 13 15 1 3 14 5 2 8 4
[41]   S4←S4, 3 15 0 6 10 1 13 8 9 4 5 11 12 7 2 14
[42]   S5← 2 12 4 1 7 10 11 6 8 5 3 15 13 0 14 9
[43]   S5←S5, 14 11 2 12 4 7 13 1 5 0 15 10 3 9 8 6
[44]   S5←S5, 4 2 1 11 10 13 7 8 15 9 12 5 6 3 0 14
[45]   S5←S5, 11 8 12 7 1 14 2 13 6 15 0 9 10 4 5 3
[46]   S6← 12 1 10 15 9 2 6 8 0 13 3 4 14 7 5 11
[47]   S6←S6, 10 15 4 2 7 12 9 5 6 1 13 14 0 11 3 8
[48]   S6←S6, 9 14 15 5 2 8 12 3 7 0 4 10 1 13 11 6
[49]   S6←S6, 4 3 2 12 9 5 15 10 11 14 1 7 6 0 8 13
[50]   S7← 4 11 2 14 15 0 8 13 3 12 9 7 5 10 6 1
[51]   S7←S7, 13 0 11 7 4 9 1 10 14 3 5 12 2 15 8 6
[52]   S7←S7, 1 4 11 13 12 3 7 14 10 15 6 8 0 5 9 2
[53]   S7←S7, 6 11 13 8 1 4 10 7 9 5 0 15 14 2 3 12
[54]   S8← 13 2 8 4 6 15 11 1 10 9 3 14 5 0 12 7
[55]   S8←S8, 1 15 13 8 10 3 7 4 12 5 6 11 0 14 9 2
[56]   S8←S8, 7 11 4 1 9 12 14 2 0 6 10 13 15 3 5 8
[57]   S8←S8, 2 1 14 7 4 10 8 13 15 12 9 0 3 5 6 11
[58]   S← 8 4 16 ⍴S1,S2,S3,S4,S5,S6,S7,S8
[59]   PC1A← 57 49 41 33 25 17 9
[60]   PC1A←PC1A, 1 58 50 42 34 26 18
[61]   PC1A←PC1A, 10 2 59 51 43 35 27
[62]   PC1A←PC1A, 19 11 3 60 52 44 36
[63]   PC1B← 63 55 47 39 31 23 15
[64]   PC1B←PC1B, 7 62 54 46 38 30 22
[65]   PC1B←PC1B, 14 6 61 53 45 37 29
[66]   PC1B←PC1B, 21 13 5 28 20 12 4
[67]   PC2← 14 17 11 24 1 5
```

Figure C Continued

```
[68]   PC2←PC2, 3 28 15 6 21 10
[69]   PC2←PC2, 23 19 12 4 26 8
[70]   PC2←PC2, 16 7 27 20 13 2
[71]   PC2←PC2, 41 52 31 37 47 55
[72]   PC2←PC2, 30 40 51 45 33 48
[73]   PC2←PC2, 44 49 39 56 34 53
[74]   PC2←PC2, 46 42 50 36 29 32
     ∇
```

Figure C Continued

INDEX